Global-Change Scenarios

Their Development and Use

Synthesis and Assessment Product 2.1b
Report by the U.S. Climate Change Science Program
and the Subcommittee on Global Change Research

AUTHORS:

Edward A. Parson, University of Michigan
Virginia R. Burkett, U.S. Geological Survey
Karen Fisher-Vanden, Dartmouth College
David W. Keith, University of Calgary
Linda O. Mearns, National Center for Atmospheric Research
Hugh M. Pitcher, Pacific Northwest National Laboratory
Cynthia E. Rosenzweig, National Aeronautics and Space Administration
Mort D. Webster, Massachusetts Institute of Technology

CLIMATE CHANGE SCIENCE PROGRAM PRODUCT DEVELOPMENT ADVISORY COMMITTEE

Eight members of the Climate Change Science Program Product Development Advisory Committee (CPDAC) wrote this Climate Change Science Program Synthesis and Assessment Product at the request of the Department of Energy. The entire CPDAC has accepted the contents of the product. Recommendations made in this report regarding programmatic and organizational changes, and the adequacy of current budgets, reflect the judgment of the report's authors and the CPDAC and are not necessarily the views of the U.S. Government.

Chair
Robert M. White, The Washington Advisory Group

Vice Chair
Soroosh Sorooshian, University of California-Irvine

Designated Federal Officer
Anjuli S. Bamzai, Department of Energy
Office of Biological and Environmental Research

Members

David C. Bader
Lawrence Livermore National Laboratory

*Virginia R. Burkett
U.S. Geological Survey

Antonio J. Busalacchi
University of Maryland

Leon E. Clarke
Pacific Northwest National Laboratory

Curtis C. Covey
Lawrence Livermore National Laboratory

James A. Edmonds
Pacific Northwest National Laboratory

*Karen Fisher-Vanden
Dartmouth College

Brian P. Flannery
Exxon-Mobil Corporation

William J. Gutowski
Iowa State University

David G. Hawkins
Natural Resources Defense Council

Isaac M. Held
Geophysical Fluid Dynamics Laboratory

Henry D. Jacoby
Massachusetts Institute of Technology

*David W. Keith
University of Calgary

Kenneth E. Kunkel
Illinois State Water Survey

Richard S. Lindzen
Massachusetts Institute of Technology

*Linda O. Mearns
National Center for Atmospheric Research

Ronald L. Miller
National Aeronautics and Space Administration

*Edward A. Parson
University of Michigan

*Hugh M. Pitcher
Pacific Northwest National Laboratory

William A. Pizer
Resources for the Future

John M. Reilly
Massachusetts Institute of Technology

Richard G. Richels
Electric Power Research Institute

*Cynthia E. Rosenzweig
National Aeronautics and Space Administration

Robin T. Tokmakian
Naval Postgraduate School

*Mort D. Webster
Massachusetts Institute of Technology

Julie A. Winkler
Michigan State University

Gary W. Yohe
Wesleyan University

Minghua H. Zhang
Stony Brook University

*Authors for *Global Change Scenarios: Their Development and Use*

July 2007

Members of Congress:

On behalf of the National Science and Technology Council, the U.S. Climate Change Science Program (CCSP) is pleased to transmit to the President and the Congress this report, *Scenarios of Greenhouse Gas Emissions and Atmospheric Concentrations and Review of Integrated Scenario Development and Application.* This is the second in a series of Synthesis and Assessment Products produced by the CCSP. This series of 21 reports is aimed at providing current evaluations of climate change science to inform public debate, policy, and operational decisions. These reports are also intended to help inform CCSP's consideration of future program priorities. This second Synthesis and Assessment Product issued pursuant to Section 106 of the Global Change Research Act of 1990 and has two components: "Development of New Scenarios of Greenhouse Gas Emissions and Atmospheric Concentrations" (Part A) and a "Review of Integrated Scenario Development and Application" (Part B). Here we transmit to you Part B.

CCSP's guiding vision is to provide the Nation and the global community with the science-based knowledge to manage the risks and opportunities of change in the climate and related environmental systems. The Synthesis and Assessment Products are important steps toward that vision, helping translate CCSP's extensive observational and research base into informational tools directly addressing key questions that are being asked of the research community.

This product will contribute to and enhance the ongoing and iterative international process of producing and refining climate-related scenarios and scenario tools. It was developed with broad scientific input and in accordance with the Guidelines for Producing CCSP Synthesis and Assessment Products, Section 515 of the Treasury and General Government Appropriations Act for Fiscal Year 2001 (Public Law 106-554), and the Information Quality Act guidelines issued by the Department of Energy pursuant to Section 515. The CCSP Interagency Committee relies on Department of Energy certifications regarding compliance with Section 515 and the Guidelines for Producing CCSP Synthesis and Assessment Products.

We commend the report's authors for both the thorough nature of their work and their adherence to an inclusive review process.

Samuel W. Bodman
Secretary of Energy

Chair, Committee on
Climate Change
Science and Technology Integration

Carlos M. Gutierrez
Secretary of Commerce

Vice-Chair, Committee on
Climate Change
Science and Technology Integration

John H. Marburger III, Ph.D.
Director, Office of
Science and Technology Policy
Executive Director, Committee on
Climate Change
Science and Technology Integration

ACKNOWLEDGEMENT

This report has been peer reviewed in draft form by individuals chosen for their diverse perspectives and technical expertise. The expert review and selection of reviewers followed the OMB's Information Quality Bulletin for Peer Review. The purpose of this independent review is to provide candid and critical comments that will assist the Climate Change Science Program in making this published report as sound as possible and to ensure that the report meets institutional standards. The peer review comments, draft manuscript, and response to the peer review comments are publicly available at: www.climatescience.gov/Library/sap/sap2-1/default.php.

We wish to thank the following individuals for their peer review of this report:

Frans Berkhout, Free University of Amsterdam

Garry Brewer, Yale University

Ged Davis, World Economic Forum

Nebojsa Nakicenovic, International Institute for Applied Systems Analysis

Robert Lempert, RAND Corporation

M. Granger Morgan, Carnegie-Mellon University

John Robinson, University of British Columbia

Thomas J. Wilbanks, Oak Ridge National Laboratory

We would also like to thank the numerous individuals who provided their comments during the public comment period. The public review comments, draft manuscript, and response to the public comments are publicly available at: www.climatescience.gov/Library/sap/sap2-1/default.php.

The writing team is grateful for research assistance by Kristen Cleary, Emily Kelly, Paul Porter, Gautam Rao, and Nora Van Horssen.

Editor ....Elizabeth L. Malone

Technical Advisor ....David Dokken

Graphic Production ....DesignConcept

Recommended Citations

The Entire Volume - CCSP Synthesis and Assessment Product 2.1
CCSP, 2007: Scenarios of Greenhouse Gas Emissions and Atmospheric Concentrations (Part A) and Review of Integrated Scenario Development and Application (Part B). A Report by the U.S. Climate Change Science Program and the Subcommittee on Global Change Research [Clarke, L., J. Edmonds, J. Jacoby, H. Pitcher, J. Reilly, R. Richels, E. Parson, V. Burkett, K. Fisher-Vanden, D. Keith, L. Mearns, C. Rosenzweig, M. Webster (Authors)]. Department of Energy, Office of Biological & Environmental Research, Washington, DC., USA, 260 pp.

This Sub-Report (2.1B)
Parson, E., V. Burkett, K. Fisher-Vanden, D. Keith, L. Mearns, H. Pitcher, C. Rosenzweig, M. Webster, 2007. *Global Change Scenarios: Their Development and Use.* Sub-report 2.1B of Synthesis and Assessment Product 2.1 by the U.S. Climate Change Science Program and the Subcommittee on Global Change Research. Department of Energy, Office of Biological & Environmental Research, Washington, DC., USA, 106 pp.

The Companion Sub-Report (2.1A)
Clarke, L., J. Edmonds, H. Jacoby, H. Pitcher, J. Reilly, R. Richels, 2007. *Scenarios of Greenhouse Gas Emissions and Atmospheric Concentrations.* Sub-report 2.1A of Synthesis and Assessment Product 2.1 by the U.S. Climate Change Science Program and the Subcommittee on Global Change Research. Department of Energy, Office of Biological & Environmental Research, Washington, DC., USA, 154 pp.

TABLE OF CONTENTS

EXECUTIVE SUMMARY

A scenario is a description of potential future conditions produced to inform decision-making under uncertainty. Scenarios can help inform decisions that involve high stakes and poorly characterized uncertainty, which may thwart other, conventional forms of analysis or decision support. Originally developed to study military and security problems, scenarios are now widely used for strategic planning and assessment in businesses and other organizations, and increasingly to inform planning, analysis, and decision-making for environmental issues, including climate change.

Scenarios can serve many purposes. They can help inform specific decisions, or can provide inputs to assessments, models, or other decision-support activities when these activities need specification of potential future conditions. They can also provide various forms of indirect decision support, such as clarifying an issue's importance, framing a decision agenda, shaking up habitual thinking, stimulating creativity, clarifying points of agreement and disagreement, identifying and engaging needed participants, or providing a structure for analysis of potential future decisions.

SCENARIOS FOR CLIMATE CHANGE: FIVE TYPES

Developing a scenario exercise involves many design choices, of which the most important involve choosing the few key uncertainties to represent in alternative scenarios. Five types of scenarios have been developed to address different aspects of the climate-change issue; these are distinguished by where they fall along a simple linear causal chain extending from the socio-economic determinants of greenhouse-gas emissions through the impacts of climate change as shown in Figure ES-1. (This figure does not represent the complete causal structure of the climate issue, which has many linkages and feedbacks. Rather, this simple structure only illustrates how scenarios have been used to fit within the simplest and most prominent causal pathway of the issue.)

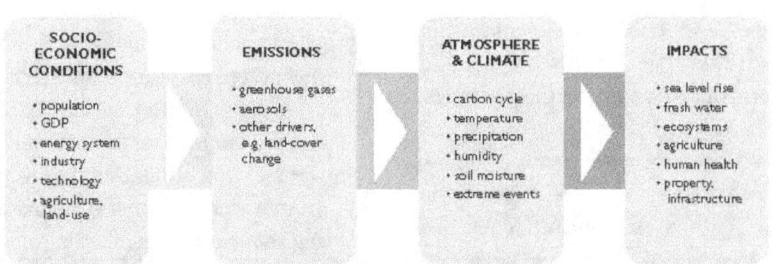

Figure ES-1. Scenarios of anthropogenic climate change: simple linear causal chain

Emissions Scenarios for Climate Simulations: Emissions scenarios present future paths of greenhouse-gas emissions or other climate perturbations. A major use of these is to provide needed inputs to climate models. Such scenarios may specify simple arbitrary perturbations of emissions or concentrations (e.g., doubling atmospheric CO_2), or time-paths reflecting specified assumptions for evolution of socio-economic drivers such as population, economic growth, and technological change.

Emissions Scenarios for Exploring Alternative Energy/Technology Futures: Another use of emissions scenarios involves specifying an environmental or emissions target, arbitrarily or based on normative or political goals, to examine what patterns of socio-economic change, energy resources, and technology development are consistent with the target and/or what interventions might be needed to meet it. Such scenarios have examined conditions for stabilizing atmospheric CO_2 concentration at various levels and the implications of stabilizing radiative forcing for multi-gas reduction strategies.

Climate-Change Scenarios: Climate scenarios specify potential future climate conditions to inform assessments of impacts, vulnerabilities, and adaptation options, and inform decision-making for adaptation or mitigation. They can be produced by arbitrary perturbation of present conditions, by using climates from elsewhere or the past as a proxy for potential future climate in a given location, or by climate-model simulations driven by some specified scenario of future emissions.

Scenarios of Direct Biophysical Impacts, e.g., Sea Level Rise: Scenarios can specify alternative trajectories for some important form of climate impact that influences many other impacts. For example, scenarios of sea level rise can capture the most important impact pathways in many coastal regions, including the large uncertainties associated with potential loss of continental ice sheets in Greenland and Antarctica.

Multivariate Scenarios for Impact Assessment: Assessing climate-change impacts requires not just considering climate in isolation, but other linked changes and stresses, including both environmental and socio-economic

trends. The factors that influence particular impacts and vulnerabilities are likely to be widely variable, and may include demographic, economic, technological, institutional, and cultural characteristics. Consequently, scenarios may have to be generated in an exploratory manner in the context of attempting to assess specific local and regional impacts.

SCENARIOS FOR CLIMATE CHANGE: MAJOR EXAMPLES

The report reviews four major exercises producing or using scenarios for climate-change applications. The examples include national and international activities, produced by different sets of actors for different purposes.

The Intergovernmental Panel on Climate Change (IPCC) has produced three sets of scenarios of 21st-century greenhouse-gas emissions, of which the most ambitious and important were produced for the *Special Report on Emissions Scenarios* (SRES) between 1997 and 1999. SRES produced four qualitative storylines on which six "marker" scenarios were based – one model quantification of each storyline plus two technological variants of one storyline that stressed fossil-intensive and low-carbon energy supply technologies – each produced by a different energy-economic model. Other models' replications of each other's marker scenarios plus a few additional explorations yielded 40 scenarios in total. These scenarios highlighted several insights, including the ability of alternative paths with similar emissions in 2100 to differ widely in their interim pathways and thus in atmospheric concentrations; the ability of alternative technological assumptions alone to generate as wide a range of emissions futures as substantially divergent socio-economic pathways; and the fact that similar emissions paths can come from widely different combinations of underlying socio-economic factors and so pose distinct mitigation problems. A widely publicized critique of the SRES scenarios alleged over-estimation of future emissions growth due to the metric used to compare incomes in rich and poor nations, but the overestimation was later found to be insignificant. More serious and illuminating challenges associated with these scenarios concerned how to balance and integrate qualitative and quantitative scenarios; how to use and how much to coordinate multiple models to generate the most useful insights; and whether, when, and how it is appropriate to assign explicit probability judgments to alternative scenarios or associated ranges of quantitative variables.

The US National Assessment was a comprehensive assessment of potential impacts of climate change and variability on the United States, focusing on major regions and sectors (agriculture, water, human health, coastal areas and marine resources, and forests). The National Assessment needed scenarios of 21st-century US climate and socio-economic changes. For climate scenarios, it relied principally on climate-model scenarios produced by the UK Hadley Centre and the Canadian Centre for Climate Modeling and Analysis, each driven by a single emissions scenario, with statistical downscaling based on detailed local conditions and present patterns of fine-scale climate variation. Other proposed types of climate scenario, including historical scenarios and inverse methods to probe for key vulnerabilities, were less used. For socio-economic scenarios, a novel approach was proposed that combined specified scenarios for a few key national-level variables such as population and economic growth, and a common process to elaborate and document additional socio-economic assumptions as needed for specific regional and sector analyses. The National Assessment was criticized for relying on just two climate-model runs and one emissions scenario, although these choices were dictated by time limits and availability of climate-model runs. Limited use was made of the socio-economic approach, principally due to time limits and communication problems.

The UK Climate Impacts Programme (UKCIP) provides common datasets, tools, and support, including scenarios, for climate-impact assessments for UK regions and sectors by researchers and stakeholders. The program produced climate scenarios in 1998 and 2002, all based on the Hadley Centre climate models, and socio-economic scenarios in 2001. The program stresses building a sustained assessment capability by acting as a motivator, resource, and light coordinator with little central

authority over separate assessments. The reliance on climate scenarios from just one family of climate models may pose risks of incomplete representation of key uncertainties.

The Millennium Ecosystem Assessment (MEA) examined the status, present trends, and longer-term challenges to the world's ecosystems, including climate change and other stresses. One of the assessment's four working groups constructed scenarios of global ecosystems to 2050 and beyond, largely independently of the group examining current status and trends. All assessment components used a common conceptual framework, which distinguished indirect drivers of ecosystem change (e.g., population and economic growth, technological change, policies and lifestyles), direct drivers (e.g., climate change, air pollution, and land-use and land-cover change), ecosystem indicators, ecosystem services, measures of human well-being, and response options. The Scenarios group applied this framework to characterizing potential ecosystem stresses in 2050, with more limited projections to 2100. The four scenarios were based on two dimensions of uncertainty: degree of globalization, and predominance of proactive vs. reactive response to ecosystem stresses. The qualitative storylines underlying these scenarios were more richly developed than in other climate-change scenario exercises. Concerns with these scenarios pertained to the degree of integration and consistency among qualitative and quantitative scenario components; risks of logical circularity within scenarios; and unexplained similarity of projected ecosystem effects among scenarios.

SCENARIOS FOR CLIMATE CHANGE: CHALLENGES AND CONTROVERSIES

Scenarios and Decisions

Scenarios can inform climate-change mitigation and adaptation decisions, but most uses so far have had relatively indirect connections to such decisions. Although there is no single global climate-change decision-maker, scenarios can inform the many decision-makers with diverse responsibilities that will affect and be affected by climate change. Three groups of decision-

makers with distinct information needs can be distinguished: mitigation policy-makers, who are mostly but not exclusively national officials; impacts and adaptation managers, including national officials and others who are responsible for particular climate-sensitive assets, resources, or interests; and energy resource and technology managers, who include owners, developers, and investors in energy resources and energy-related capital stock and new technologies.

A key issue in creating scenarios for all decision-makers is how to represent decisions within scenarios. In general, decisions by the scenario user should be explicitly examined relative to baseline conditions specified in scenarios, while decisions by others outside their control should be treated like any exogenous uncertainty. The issue is most important in the treatment of mitigation decisions: scenarios to inform mitigation should allow explicit examination of the entire relevant range of mitigation decisions, while scenarios to inform impacts and adaptation should specify the likely range of mitigation efforts – usually yielding a narrower range of emissions futures than is considered in scenarios to inform mitigation.

Scenarios in Assessments and Policy Debates

In climate-change assessments, scenarios can provide required inputs to other parts of the analysis and help to organize multiple components of the assessment. When scenarios are used in a prominent assessment, they may subsequently be adopted in planning or decision-support processes outside the original assessment. Scenarios can also help frame public and policy debate, in part by providing an aggregate metric of the issue's severity. They consequently may gain prominence in contentious policy debates, and so become subject to political attempts to influence their content and political criticism based on their perceived implications for policy action. The unavoidable judgments underlying construction of scenarios provide opportunity for partisan efforts to make scenarios policy prescriptive, and for claims that only certain scenarios are plausible (e.g., high- or low-emissions scenarios, depending on the critic's motivation). These claims are unavoidable, since scenarios represent key uncertainties

bearing on high-stakes policy decisions, but such attempts to restrict scenarios should be resisted, principally through prominent communication of the reasoning, assumptions, and treatment of particular uncertainties underlying scenarios.

Scenario Development Process: Expert-Stakeholder Interactions

Scenario developers must decide how and how much to involve scenario users and stakeholders in scenario development. In other fields – where users are clearly identified – relatively few and homogeneous, intensive collaboration between scenario developers and users or their representatives is desirable. Close user involvement is also advantageous in developing scenarios for climate change, but potential users of these scenarios are more numerous and diverse, may not be clearly identified, and may have contending material interests in the scenarios' content or use. This situation calls for delicate decisions about participation and representation to keep scenarios tuned to practical users' needs while keeping the development process small enough to be manageable.

Communication of Scenarios

Climate change scenarios must be communicated to multiple audiences with diverse interests and information needs. In addition to the scenarios' content, sufficient information must be provided about the process and reasoning by which the scenarios were developed, to allow users to scrutinize the underlying data, models, and reasoning; judge their confidence in the scenarios; and have opportunities to critique the scenarios and suggest alternative approaches. Effective communication can help engage a broad user community in updating and improving scenarios. Open communication of the decisions, assumptions, and uncertainties underlying scenarios is likely to both increase users' confidence that the scenarios have reasonably represented current knowledge and key uncertainties, and help them develop alternatives if they are unconvinced.

Consistency and Integration in Scenarios

Scenario developers should strive for internal consistency. At one level, this means avoiding clear contradictions with well-established knowledge and not moving inadvertently outside bounds of historical experience – although such sharp departures from experience may be useful if pursued intentionally to examine low-probability risks or broaden decision-makers' perceptions. Perceptions of internal consistency or coherence in scenarios ultimately rest on subjective judgments, which pose well-known risks of bias if not carefully structured and controlled. Potential inconsistencies grow when scenario exercises use multiple models and attempt to harmonize them, particularly when some key quantities are externally specified for some models and calculated within others. Attempting to avoid such inconsistency by standardizing model outputs, however, can carry more serious risks by obscuring interpretation of results and precluding use of model variation to illuminate uncertainty. Attempts to connect qualitative and quantitative aspects of scenarios have been particularly challenging for pursuit of consistency. Different narrative scenarios often reflect different assumptions about how the world works, which correspond more closely to different model structures than to parameter variation. Better integrating the two approaches will require developing ways to connect narrative scenarios to model structures, rather than merely to target values for a few variables that models are then asked to reproduce.

Treatment of Uncertainty in Scenarios

A scenario exercise can represent a few key uncertainties by variation among scenarios. Extreme economy is required in choosing which uncertainties to represent, what variation (including potential extremes) to represent for each, and how to combine them in a manageable number of scenarios. Complex narrative scenarios pose special problems in representing and communicating uncertainty, usually addressed by seeking underlying structural uncertainties – e.g., deep societal trends such as globalization or values shifts – that are judged to influence many other factors of concern. The most promi-

nent controversy in treatment of uncertainty in scenarios has concerned whether or not to explicitly assign probabilities to scenarios or associated ranges of quantitative outcome variables. The debate rests in part on different views of the typical contents of scenarios, since subjective probabilities can readily be assigned to ranges of one or two quantitative variables. Explicit probability assignment in such simple cases offers clear benefits for assessing alternative choices and avoids the risk of users assigning their own, perhaps less informed, probability judgments. Assigning probabilities to rich multivariate scenarios, particularly if these include narrative elements, is much more problematic, since there is no clearly defined interval "between" such scenarios and their boundaries are not clearly defined.

CONCLUSIONS AND RECOMMENDATIONS

Use of Scenarios in Climate-Change Decisions

- Scenarios can make valuable contributions to climate-change decision-making. There is a big gap between the use of scenarios in current practice and their potential contributions, but interest in using scenarios is increasing.

- Scenarios of global emissions and resultant climate change are required by many diverse climate-related decision-makers, but beyond these common requirements decision-makers' needs from climate-change scenarios are highly diverse.

- Impacts and adaptation managers include both national officials and others responsible for more specific domains of impact. They need climate-change scenarios, driven by specified global emissions scenarios, to represent potential climate-related stresses on their areas of responsibility, plus other environmental and socio-economic information at appropriate scales. Their combined needs – for centrally produced climate scenario information, associated tools and support, and a capability to develop and apply additional scenario information related to their responsibilities – suggest the need for a cross-scale organizational structure to provide scenario information.

- Mitigation policy-makers, who are mainly but not exclusively national officials, need scenarios of global and national emissions trends, resultant climate change, and aggregate impacts. In addition, they need scenario information about the potential policy environment for their choices, including alternative scenarios of other nations' mitigation strategies, international mitigation decisions, and implementation and compliance. In some cases, they can usefully employ target-driven scenarios for backcasting analysis. Mitigation decisions require scenario development capacity at the national level.

- Scenarios for mitigation decisions should include a wide range of baseline emissions assumptions and should not pre-judge the likely level of mitigation effort, while scenarios for impact and adaptation managers should be based on emissions assumptions that include the range of mitigation interventions they judge likely.

- Energy resource and technology managers, who are mainly private-sector actors, primarily need scenarios that represent alternative policy regimes over the 30- to 50-year time horizons relevant for investment and technology-development decisions. Scenarios of emissions and climate change may provide background, but do not capture the most important uncertainties for these decision-makers.

Use of Scenarios in Climate-Change Assessments

- Large-scale, official assessments are currently the main users of scenarios and will likely remain major users. Scenarios in assessments mostly support further analysis, modeling, and assessment. They can also help frame the climate issue for the public and policy-makers. Presentation of scenarios in assessments leads to additional unforeseen uses.

- Scenarios contain unavoidable elements of judgment in their production and use. This makes them vulnerable both to attempts at bias and to partisan attack. The most productive response lies in transparency about the process, reasoning, and assumptions used to produce scenarios, which can both help limit

bias in scenario production and focus subsequent argument on underlying uncertainties.

What Should Centrally Provided Emissions and Climate Scenarios Look Like?

- Centrally provided scenarios of emissions and resultant climate change should be global in scope, with major climate-relevant emissions and other perturbations specified at least for major world regions. They should have a time horizon of a century or longer, with interim results at roughly decadal resolution.

- Centrally provided scenarios of global emissions and climate change can help inform mitigation and adaptation decisions at national and sub-national scales, but such decisions require additional information at these scales.

- Emissions scenarios of several types are needed to serve diverse uses, including alternative baselines, alternative levels of incremental stringency of mitigation effort, and specified future targets to support backcasting and feasibility analysis. Some emissions scenarios should be coupled to explicit scenarios of wide-ranging alternative socio-economic futures, but this is not necessary for all uses. Scenarios should reflect various explicit degrees of coordination, including simple fully standardized scenarios for evaluating and comparing downstream models, multi-model scenarios using common input assumptions, and non-standardized scenarios to explore alternative assumptions or meet specific user needs.

- Some scenarios of socio-economic conditions should include qualitative and quantitative elements and sustained analytic efforts to link the two. These elements can provide a vehicle to explore major historical uncertainties with large implications for climate change and vulnerability; provide a logical structure to connect assumed trajectories for multiple variables; and provide guidance to other analysts or users to extend scenarios by elaborating additional detail. Alternative qualitative and narrative elements should be linked not just to alternative parameter values in quantitative models, but also to alternative forms of causal relations and model structures.

Scenario Process: Developer-User Interactions

- There is value in close collaboration between scenario developers and users, particularly at the beginning and ending stages of a scenario exercise.

- The ease of achieving such collaboration and its value are likely to be greater when scenario users are clearly identified, few in number, and similar in their interests and perspectives.

Communication of Scenarios

- Effective communication of scenarios is essential, in forms useful to audiences of diverse interests and technical skills. In addition to scenario contents, communication should include associated documentation, tools, and support.

- Transparency of underlying reasoning, assumptions, and major uncertainties is crucial. Such transparency is necessary to support the credibility of scenarios, to alert potential users to conditions under which they might wish to use or modify them, and to inform criticism and improvement of scenarios.

Consistency and Integration in Scenarios

- Any scenario should be internally consistent in its assumptions and reasoning, to the extent this can be established.

- In scenario exercises that use multiple models to explore potential uncertainties in future conditions, consistency among models should be pursued primarily through coordination of inputs, not outputs, except when coordinated outputs represent common goals for policy evaluation.

- Transparency in reporting scenario and model differences as well as underlying assumptions and reasoning can help mitigate the effects of inconsistencies among scenarios.

Treatment of Uncertainty in Scenarios

- More explicit characterization of probability judgments should be included in some future scenario exercises than has been practiced so far. Means available to express these judgments are of widely varying specificity, ranging from agreed terminology to explicitly quantified probability distributions. All such judgments should include explicit acknowledgement of their inevitable subjective elements and appropriate caveats.

- Explicit probability judgments are easiest to produce and least controversial in scenarios generated using quantitative models of climate change or specific impact domains. These can be conditioned on specific assumptions for socio-economic inputs such as emissions, and can represent explicitly and quantitatively the effects of specified variation in initial conditions or unknown parameter values. These devices are also available, although in less widespread use, in economic models used to project emissions.

- Including explicit probability judgments is likely to be most useful when key variables are few, quantitative outcomes are needed, and potential users are numerous and diverse. It is likely to be least useful when scenarios specify multiple characteristics, including prominent qualitative elements; when the purpose is sensitivity analysis or heuristic exploration; and when potential users are few, similar, and known.

- Because of their large and diverse set of potential users, centrally provided scenarios of global emissions and climate change should attempt to include some explicit probability judgments for ranges of key quantitative outputs, including global emissions and global-average temperature change. These should span a wide range of judged uncertainty on these variables, e.g., 95 to 99 percent. Providing such explicit likelihood statements lets users choose whether to use them or not.

- Scenario exercises should give more attention to low-probability, high-consequence extreme cases, such as loss of a major continental ice sheet or changes in meridional ocean circulation. With these, it is especially crucial to be explicit and transparent about the reasoning and assumptions underlying each scenario, including developers' judgments of relative likelihoods.

Expanding and Sustaining Capacity for Production and Use of Scenarios

- Present scenario capacity is inadequate. To help fulfill these presently unmet needs, the CCSP should establish a program to:

  - Commission scenarios for use in assessments and decision-support activities.

  - Disseminate scenarios with associated documentation, tools, and guidance materials.

  - Commission various groups to evaluate scenarios and their applications, and to develop improved methods.

  - Archive results and documentation related to all these tasks, to provide historical perspective and institutional memory for future scenario-related activities.

- Design and management conditions of this new program should include six elements.

  - The program should build and maintain strong connections with outside relevant

expertise, and analytic and modeling capability.

- The program should integrate and balance goals and criteria related to scientific and technical quality, and those related to utility and relevance to users.

- The program should be insulated from political control.

- The program should strive for maximum transparency in its own activities, in addition to demanding it from activities it supports.

- The program will require the authority and resources necessary to articulate and promulgate standards for transparency, consistency, and quality control.

- The program will require adequate sustained resources level of effort.

Introduction

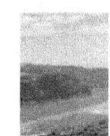

This report examines the development and use of scenarios in global climate change applications. It considers scenarios of various types – including but not limited to emissions scenarios – and reviews how they have been developed, what uses they have served, what consistent challenges they have faced, what controversies they have raised, and how their development and use might be made more effective. The report is Synthesis & Assessment Product 2.1b of the US Climate Change Science Program. By synthesizing available literature and critically reviewing past experience, the report seeks to assist those who may be conducting, using, or commissioning scenarios related to global climate change.

Scenarios are used to support planning and decision-making when issues have deep or poorly characterized uncertainty and high stakes, often accompanied by long time horizons. These conditions apply to the major decisions about how to respond to global climate change. As scientific research advances our knowledge of the climate's present state and trends, its patterns of variability, and its responses to external forcings, we are gaining an increasingly clear view of risks that may be realized late this century or beyond. These future risks are linked to near-term socio-economic trends and decisions in both public and private sectors. Some near-term decisions – such as investment in long-lived capital equipment, new resources, or new technologies in the energy sector – can influence long-term trends in the emissions contributing to climate change. Other near-term decisions – such as investment in water resources infrastructure or coastal development – can influence how adaptable and how vulnerable future society will be to the impacts of climate change.

Although such decisions are being made now, making them responsibly requires considering their potential consequences over the longer term, including associated uncertainties. This requires thinking about the future conditions that will shape their consequences, not just next month or next year but 10, 30, 50, or 100 years in the future – longer periods than the horizon of conventional methods of planning or analysis.[2] Attempting to describe potential future conditions over this long time horizon presents a seeming paradox. On the one hand, conditions this far in the future, and the factors and actors that may influence them, are deeply uncertain.[3] On the other hand, we have a great deal of knowledge that can help make informed assumptions about future conditions, even over such long horizons. This includes well-established scientific knowledge about physical, chemical, and biological processes; more weakly, relatively well-established causal mechanisms in economics, sociology, and politics; and more weakly still, certain seemingly robust empirical patterns of historical change in population, economics, and technology. All of these give some guidance to support judgments about future conditions that are more or less likely, virtually certain, or virtually impossible. In some ways we might be highly confident that the future will resemble the present, e.g., in the radiative properties of atmospheric trace gases. In others, we might judge it likely that future conditions will lie within some envelope extrapolated from present and past trends, e.g., in projecting rates of change in fertility, mortality, or labor productivity. Still other areas, such as the development and social consequences of major technological advances, or large-scale political events such as wars, political realignments, or epidemics, may hold more fundamental uncertainties. In some cases, such uncertainties may

be adequately represented as wider distributions of recognized uncertain quantities. In others, they may represent events whose character or even possibility we have not yet imagined.

Despite pervasive uncertainties, people must make near-term decisions related to climate change that have long-term consequences, including potential irreversibilities. Scenarios are tools to help inform these decisions by gathering and organizing relevant knowledge, organizing associated uncertainties, and structuring and disciplining associated speculation. This report assesses experience to date in developing and using scenarios for global climate change.

Early climate-change debates mainly concerned scientific questions such as whether and how the climate is changing, how much change is caused by human activities, and how sensitive the climate system is. Scenarios did not figure prominently in these debates. But as advancing climate science has increasingly shifted the debate from confirming and describing the climate-change problem toward deciding what to do about it, the need for long-term decision-support tools like scenarios has increased, as have the scrutiny and criticism these have attracted.[4] In a contentious public-policy area like climate change, controversy over scenarios is to be expected: scenarios are a method to structure and communicate the most important uncertainties, and conflicting judgments about uncertainties are a major source of disagreements over what to do. Consequently, we expect the trend of scenarios' increasing prominence and contentiousness to continue – particularly for emissions scenarios, since these are the relevant metric of human environmental burden and the point of most contested proposed intervention.

---

2   Morgan et al. 1998.

3   Lempert et al. 2006.

4   See, e.g., Lomborg 2001; Michaels 2003b; Castles and Henderson 2003a, b; UK House of Lords 2005.

In this report, we try to cast some light on current and coming debates over climate-change scenarios. These debates presently exhibit basic confusion about the definition, purposes, and potential uses of scenarios. We aim to provide clarification and practical advice to two related audiences: those conducting assessments or analyses that develop or use scenarios; and those commissioning, using, and interpreting such assessments or analyses. For the first group, we seek to provide an organized summary of relevant experience in similar efforts, discussion and clarification of key choices and challenges, and – to the extent present knowledge allows – practical guidance about pitfalls, challenges, and opportunities in particular approaches. For the second group, we seek to provide guidance on what to ask for, how and how much to participate in its production, how to interpret the results, and what questions to ask.

Because the charge of this report is unlike those of other Synthesis and Assessment products, the approach we have taken to producing it is necessarily different as well. We were tasked with reviewing, interpreting, and evaluating experience with scenario methods in global climate change applications. This is not a narrowly focused question, and there is not a well-developed scientific literature on which we can draw for answers. While we have reviewed the existing literature on scenarios, most of it concerns scenarios in other decision domains than global climate change. In addition, we have examined several major scenario exercises in global-change applications. In this, we have drawn on published materials, both from the exercises themselves and from commentary and criticism, as well as documentary materials and records, interviews with participants and users, and the experience and judgments of team members.

Our review of this experience has not been entirely independent, since members of this writing team were involved as participants, reviewers, and critics in two of the scenario exercises we review, the IPCC SRES process and the US National Assessment. While we have drawn on the experience of these team members, we have attempted to limit the risk of idiosyncratic interpretations and bias by drawing on other sources as well and by engaging all team members in developing our summary and discussions of these exercises. Moreover, our purpose is not to either attack or defend these past exercises, but to seek to understand the decisions made in conducting them, the factors that influenced them, and the constraints under which they operated, in order to assess their experience, identify successes and pitfalls, and to the extent possible, provide guidance to advance scenario methods for climate change and other similar environmental issues. Because the experience we review does not provide a sufficiently large or random sample to support strong scientific inference, the diagnoses, interpretations, and recommendations we present rely on our collective judgment. We have endeavored to follow our own advice, and be as transparent as possible about the foundation and reasoning underlying our conclusions and recommendations.

The report is organized as follows. Drawing on the broader literature on scenarios – most of which concerns domains other than climate change – Section 1 introduces scenarios, sharpens their definition, and outlines a few major dimensions of variation and decisions that must be made in developing a scenario exercise. Section 2 focuses specifically on scenarios for global climate change, and outlines the types of decisions that could use scenarios and the main types of scenarios that have been developed for this issue. Section 3 reviews four major expe-

We aim to provide clarification and practical advice to two related audiences: those conducting assessments or analyses that use scenarios; and those commissioning, using, and interpreting such assessments or analyses.

riences in developing and using global-change scenarios. Section 4 discusses several issues that have posed key challenges in climate-change scenarios and that are likely to require particular attention in designing new scenario exercises. In addition to drawing on Section 3, this discussion also makes use of briefer discussions of eight other examples of global-change scenarios that illustrate particular issues or challenges; these examples are presented as short boxes in Section 4. Section 5 provides our conclusions and recommendations for future development and use of global climate-change scenarios.

Scenarios, Their Characteristics and Uses

1.1 DEFINING SCENARIOS

A scenario is a description of potential future conditions, developed to inform decision-making under uncertainty. The decisions in question can be faced by individuals, groups, organizations, or governments, and may pertain to any subject matter. While many writers on scenarios give no explicit definition, others have offered a wide range of definitions, often substantially more complex and restrictive than this one. The published definitions gathered in Box 1.1 give a sense of both the broad commonalities among many analysts' conceptions and the significant differences among them.

BOX 1.1 Scenarios: a Sampling of Published Definitions.

A scenario is a coherent, internally consistent, and plausible description of a possible future state of the world.[5]

A scenario is a story that describes a possible future. It identifies some significant events, the main actors and their motivations, and it conveys how the world functions. Building and using scenarios can help people explore what the future might look like and the likely challenges of living in it.[6]

Scenarios are images of the future, or alternative futures. They are neither predictions nor forecasts. Rather, each scenario is one alternative image of how the future might unfold. A set of scenarios assists in the understanding of possible future developments of complex systems. Some systems, those that are well understood and for which complete information is available, can be modeled with some certainty, as is frequently the case in the physical sciences, and their future states predicted. However, many physical and social systems are poorly understood, and information on the relevant variables is so incomplete that they can be appreciated only through intuition and are best communicated by images and stories. Prediction is not possible in such cases.[7]

continued on next page

---

[5] IPCC 2001b:149.

[6] Shell International 2003.

[7] Nakicenovic and Swart 2000:62.

BOX 1.1, continued from previous page.

A climate scenario is a plausible representation of future climate that has been constructed for explicit use in investigating the potential impacts of anthropogenic climate change. Climate scenarios often make use of climate projections (descriptions of the modeled response of the climate system to scenarios of greenhouse gas and aerosol concentrations), by manipulating model outputs and combining them with observed climate data.[8]

[Scenarios] are created as internally consistent and challenging descriptions of possible futures. They are intended to be representative of the ranges of possible future developments and outcomes in the external world. What happens in them is essentially outside our own control.[9]

Scenarios are coherent, internally consistent and plausible descriptions of possible future states of the world, used to inform future trends, potential decisions, or consequences. They can be considered as a convenient way of visioning a range of possible futures, constructing worlds outside the normal timespans and processes covering the public policy environment.[10]

Scenarios are plausible, challenging, and relevant sets of stories about how the future might unfold. They are generally developed to help decision-makers understand the wide range of potential futures, confront critical uncertainties, and understand how decisions made now may play out in the future. They are intended to widen perspectives and illuminate key issues that might otherwise be missed or dismissed. The goal of developing scenarios is often to support more informed and rational decision-making that takes both the known and the unknown into account.[11]

The historical uses of scenarios for planning and analysis lie in war games, exercises of simulated conflict used for military training, planning, and operational decision-making. The roots of war games extend to antiquity, although the first formalized war games were developed for officer training in 19th-century Prussia.[12] In the 1940s and 1950s, exercises resembling war games began to be applied outside the purely military domain, to study potential international crises that included both high-level political decision-making and the potential for military conflict. These exercises were informed by the then-new field of game theory, which promised new formal insights into situations of conflict and strategic decision-making,[13] motivated by the recognition that the new nuclear age had both raised the stakes of international diplomacy and created profound new uncertainties over how to proceed. In these exercises, principally devel-

oped at the Rand Corporation, scenarios were sketches of challenging but plausible situations to which participants had to respond, allowing exploration of associated threats and opportunities. They adopted the term "scenario" from film and theatre, where it denotes a brief sketch of a story that includes only enough detail to convey broad points of plot and character. As in classic war games, scenarios in these exercises served to help organizations and their leaders prepare for novel, complex challenges that they might not anticipate, and which – if they did arise – would likely develop too fast to allow adequate reflection or analysis in real time.[14]

Over the past few decades, the use of scenarios has moved outside the realm of military and diplomatic activity. Beginning with strategic planning at the Royal Dutch/Shell oil company,[15] scenarios are now widely used for strategic planning, analysis, and assessment by

[8] IPCC 2001a:741.

[9] van der Heijden 1996:5.

[10] Berkhout et al. 2001:i.

[11] MEA 2006:xvii.

[12] Brewer and Shubik 1979.

[13] Von Neumann and Morgenstern 1944, Nash 1950.

[14] Levine 1964a,b; Schelling 1964; DeWeerd 1967, 1975; Brewer and Shubik 1979.

[15] For relevant history, see Hausrath 1971, Shubik 1975, Greenberger et al. 1983, Huss 1988, Schoemaker 1995, Schultz and Sullivan 1972, Schwartz 1991, Shell International 2003.

businesses and other organizations. They have also figured increasingly prominently in planning, analysis, and policy debate for long-term environmental issues, in particular global climate change. Because the total body of experience with scenarios provides useful insights into their use in any particular domain, this section elaborates the meaning, characteristics, and potential uses of scenarios in general. The next section turns to their specific use for global environmental issues.

1.1.1 Distinguishing scenarios from assessments, models, and analyses

Confusion is widespread in discussions of scenarios, because their form and usage are highly diverse, and because writers' uses of the term are often imprecise and occasionally contradictory. Scenarios must be distinguished, on the one hand, from assessments and various types of decision support activity that often use scenarios; and, on the other hand, from other types of statements about future conditions, such as predictions, projections, or forecasts.

An assessment is any process that reviews and synthesizes scientific or other expert knowledge to provide information of relevance to policy- or decision-makers.[16] The most common methods of assessment are deliberations of expert panels and formal models, but other methods combine human deliberations with formal analysis or modeling, including war games or other simulation games, policy exercises, political-military exercises, constructing future histories, backcasting, and others.[17] These methods may use specifications of potential future conditions – i.e., scenarios – as inputs to or components of their work. Scenarios may even be essential for some of these methods. For example, a war or crisis gaming exercise needs a scenario to specify the nature of the threat or crisis, while a formal model used to represent future development of some issue of concern needs a scenario to specify future values of those inputs not explicitly calculated within the model. But these methods are broader than and distinct from

scenarios. For example, models do not need scenario-based inputs when used to reconstruct past conditions or study causal processes.

The distinction between assessments and scenarios is perhaps clearest in conventional assessments based on deliberations of expert panels, such as the IPCC, US National Assessment, or Millennium Ecosystem Assessment (MEA). Such assessments often construct representations of future development of an issue, usually based on formal models. These representations require scenario-based inputs and may produce outputs that are themselves used as scenarios in other activities. But the scenario-related activities are frequently not the central focus of the overall assessment, which may examine many additional things, e.g., the state of knowledge in particular scientific areas, the status of and trends in particular environmental conditions, the evidence attributing particular environmental changes to human inputs, or particular policy-relevant scientific questions. Assessments may also include evaluations of proposed actions or proposed criteria for conducting such evaluations. Scenarios thus may provide required inputs to assessments, but are distinct from them.

1.1.2 Distinguishing scenarios from projections, predictions, and forecasts

Scenarios must also be distinguished from other types of statements about the future, such as predictions, projections, and forecasts. All of these satisfy the basic definition above: they are descriptions of potential future conditions whose primary purpose in most cases is to support decisions. How can scenarios be distinguished? Examining the ways scenarios are used and discussed by practitioners and researchers suggests four characteristics that distinguish them from other types of future statements. Although these characteristics are not essential, they are all more likely to be present in scenarios than in other types of future statement, so they help to sharpen and delimit the definition of a scenario.

First, scenarios are multi-dimensional: they describe multiple characteristics that collectively make up a coherent representation of future

---

16   Parson 2003:9; Mitchell et al. 2006.

17   NRC 1996; Hausrath 1971; Brewer 1986; Shubik 1975; Svedin and Aniansson 1987; Schultz and Sullivan 1972; Jones 1985; Parson 1996, 1997.

Effective scenarios integrate their diverse elements in a way that is coherent, communicates a clear theme or organizing principle, and avoids internal contradiction.

conditions. To achieve this, scenarios assemble and organize available knowledge, information, and assumptions from diverse bodies of research and expert judgment. The elements of a scenario can be of diverse types: quantitative or qualitative, precise or fuzzy, based on well-established research or informed speculation. Effective scenarios integrate their diverse elements in a way that is coherent, communicates a clear theme or organizing principle, and avoids internal contradiction.

Second, scenarios are schematic: that is, they are multidimensional but not without limit. Scenarios do not seek to describe potential future conditions with complete precision or detail. Rather, they highlight essential characteristics and processes with enough detail that knowledgeable observers perceive them as realistic and relevant, but not so much as to distract from large-scale patterns. Indeed, one potential use of scenarios is to stimulate creative thinking and insights, for which they must leave something to the imagination. How much detail and precision is appropriate is a judgment that depends on the particular application.

Third, scenarios usually come in groups. To be a useful tool to inform decision-making under uncertainty, scenarios must represent uncertainty. This is most often done by providing multiple scenarios, each presenting an alternative realization of uncertain future conditions.[18] The number of scenarios depends on the application. Scenario exercises usually use between two and seven, depending on the stakes of the issue, the resources invested in the exercise, and the depth of analysis devoted to each scenario. The most frequently proposed numbers are three or four.

Finally, scenarios tend to claim less confidence than other types of future statements. Although different authors' usage is not consistent, "prediction" and "forecast" usually denote statements for which the highest confidence is claimed. "Projection" denotes a less confident statement, which may have some specified con-

fidence level and may be explicitly contingent on specified assumptions about other future conditions. Calling a future statement a "scenario" usually implies still less confidence and more associated contingencies. Any use of a scenario for serious planning or analysis does, however, presume some minimal threshold of likelihood. The situation described must be judged likely enough to merit attention, and to justify expending resources and effort to study its implications and potential responses to it. There may also be a time ordering among these three types of statements – predictions or forecasts tend to describe nearer-term futures and scenarios longer-term futures – but there are exceptions, and the meaning of near term and long term depend on the particular context.

1.2. CREATING A SCENARIO EXERCISE: KEY CHARACTERISTICS AND CHOICES

Beyond these general characteristics, scenarios vary greatly in their use, production, and contents. Extensive scholarly effort has gone into providing alternative scenario taxonomies.[19] Scenarios can be distinguished, for example, by whether they present a snapshot of a future state or a dynamic account of changes over time to reach that state; by their degree of complexity; by the relative balance of deliberation and intuition versus formal analysis used in producing them; or by their temporal and spatial scale. The set of characteristics on which scenarios could be sorted is long and open-ended, so we make no attempt to provide an exhaustive list. Instead, we summarize the main dimensions of scenario variation in the form of a list of potentially open-ended design choices.

1.2.1 Variation among scenarios: three basic dimensions

Three dimensions of variation, concerned with the purpose of a scenario exercise, have far-reaching implications for its design and use and so merit separate discussion. First, the intended use of a scenario exercise can vary from the more predictive to the more exploratory or

---

[18] Crisis-response exercises are often an exception, presenting one scenario at a time showing a novel challenge, to which participants must respond and which is implicitly contrasted to the status quo.

[19] See, e.g., Duncan and Wack 1994, Godet and Roubelat 1996, van Notten et al. 2003.

heuristic. It is of course a fundamental error to take a scenario's illustrative description of potential future conditions as a confident prediction of what will actually happen – in our terminology, to take a scenario as a projection or even a prediction.[20] Still, a scenario must be judged likely enough to merit the attention of busy people. Exploratory uses of scenarios may presume only this low threshold, yet have great value. For example, scenarios can be used to probe and challenge the mental models, thought habits, and unrecognized assumptions of decision-makers; to clarify points of agreement and disagreement; to identify and engage needed participants; to provide a preliminary structure for advance analysis of potential future decisions; or to seek insights into unrecognized opportunities, risks, causal linkages, or uncertainties.[21] Such insights can arise not just from examination of uncertainties, but also from meticulous critical examination of future factors that are essentially certain (e.g., strongly determined demographic trends such as the aging of industrialized-country populations) or even of present conditions whose significance has been overlooked.[22] Still, the predictive confidence accorded to scenarios is a matter of degree: carefully developed scenarios that are judged to have captured the most important uncertainties may claim some moderate degree of confidence, and reasonable distinctions may be drawn between scenarios that represent conventional versus surprising futures, best and worst cases, etc.

A related dimension of variation among scenario exercises is their proximity to decision-making.[23] Scenario exercises may involve actual decision-makers and seek to directly advise a specific, identified, near-term decision, but more frequently their relationship to concrete decisions and decision-makers is indirect. Scenarios may be used for risk assessment, contingency planning, identification of potential threats or actions to be considered, or early characterization of a poorly understood issue. In such applications, exploratory uses dominate. Scenario exercises that are closer to decisions with significant stakes operate under very different requirements, usually driven by specific user needs. Their uses tend to be more predictive – constrained by limits of available knowledge – so they might be expected, for example, to provide more explicit and complete characterization of major uncertainties. They are also likely to be more integrated with methods to evaluate alternative choices and identify preferred ones.

A third basic dimension of variation concerns whether scenarios are defined primarily normatively, on the basis of their perceived desirability or undesirability, or primarily on the basis of their perceived plausibility or likelihood. Although all scenarios include both positive and normative elements, it is important to keep as clear as possible which elements are included based on perceived likelihood or plausibility and which because of perceived desirability or undesirability. The most frequent use of explicitly normative scenarios involves constructing some hypothetical future state primarily on the basis of its desirability. Such a future might be constructed to embody participants' general intuitions about desirable social trends, or to achieve specific environmental, development, or other goals.[24] The scenario exercise then consists primarily of backcasting – attempting to construct paths that connect present conditions to the specified future conditions, to examine the feasibility of the target, and identify costs, tradeoffs, and conditions associated with meeting it.[25] Similarly, one can posit an undesirable future state and then reason through conditions associated with avoiding it. This approach is

Scenarios can be used to probe and challenge the mental models, thought habits, and unrecognized assumptions of decision-makers; to clarify points of agreement and disagreement; to identify and engage needed participants; to provide a preliminary structure for advanced analysis of potential future decisions; or to seek insights into unrecognized opportunities, risks, causal linkages, or uncertainties.

---

[20]   Several such errors are collected and discussed in Bracken 1977, 1990; and Brewer 1990.

[21]   Brewer 1990.

[22]   Shell International 2001, 2003. For example, in a 1960s crisis exercise on a Soviet invasion of Iran, one participant realized the local supply of jet fuel available to support a rapid US response was ten times larger than had been thought, because kerosene – an acceptable substitute – was used for domestic cooking and heating (Schelling 1994).

[23]   This dimension is presented by Van Notten et al. 2003 as "exploration" versus "decision support."

[24]   See, for example, the simple scenario exercise in NRC (1999:161-176) that posited specific targets to reduce world hunger and greenhouse-gas emissions by year 2050, or the scenarios of the Global Scenario Group, which included some defined by specified trends and others back-cast from normatively specified targets for 2050 (Kemp-Benedict et al. 2002, Raskin et al. 2002).

[25]   Robinson 1982, 2003.

sometimes proposed to reduce the risks of hidden bias in construction of scenarios which, like any decision support tool, can be misused to provide support for a decision already made for other reasons, rather than to inform a decision not yet made. By bundling normative assumptions into the future target state or boundary conditions, analysts hope to reduce their penetration into the subsequent instrumental reasoning about actions and conditions to reach the target.[26]

1.2.2 Developing scenarios: main dimensions of choice

Table 1.1 extends the preceding discussion, summarizing the main areas of variation and choice involved in constructing a scenario exercise. This is a highly simplified representation of a complex set of choices. In any particular scenario exercise some of these choices may be made by default, without explicit consideration, perhaps because the preferred choice is obvious in context. Although we present these choices in simple sequential order, this is not necessary: choices might be made in some other order, or iteratively adjusted. But while the process and sequence of choices may be idealized, the set of choices is not: creating a scenario requires explicit or implicit choices on all these dimensions.

> *But whether the relationship of a scenario exercise to decisions is near or far, direct or indirect, clear understanding of its focus and purpose is important, and infrequently achieved.*

The most basic choices in developing scenarios, which include the three dimensions of variation called out above, involve identifying the main focus of the exercise: what issues are to be addressed or what decisions informed, for whom? The decision to conduct a scenario-based exercise does not necessarily imply that these matters are clearly understood. The closer a scenario exercise is to concrete decisions, the more likely it is that these definitional issues will be understood clearly, in part through discipline on the process imposed by the involvement of decision-makers. But most often, the coupling of scenarios to decisions is relatively weak.[27] In some applications (e.g., corporate strategic planning, responding to a novel military threat) the relevant decision-makers may be clearly identified at the outset, but the issues to be addressed and relevant choices may not be. In other applications, scenarios may be developed to address some broad issue or concern (e.g., climate change, emerging infectious diseases, or terrorism), but the potential users and decisions to be informed might both be unspecified. But whether the relationship of a scenario exercise to decisions is near or far, direct or indirect, clear understanding of its focus and purpose is important, and infrequently achieved: many scenario exercises muddle through with vagueness, confusion, or disagreement regarding the focus, purpose, and intended user of the exercise. Clarifying the overall focus of a scenario exercise may require broad consultations or scoping workshops involving many potentially interested decision-makers, other stakeholders, and analysts and researchers.

A second basic set of decisions concerns the process by which scenarios are developed. Like the focus of the exercise, decisions about the process of developing scenarios often receive little thought, or are not even explicitly recognized as choices, but are nonetheless highly consequential. What expertise must be included to ensure the scenarios adequately reflect the best available scientific knowledge, data, and models? What decision-makers, stakeholders, or their surrogates must be involved to keep the scenarios relevant, plausible, and credible? For

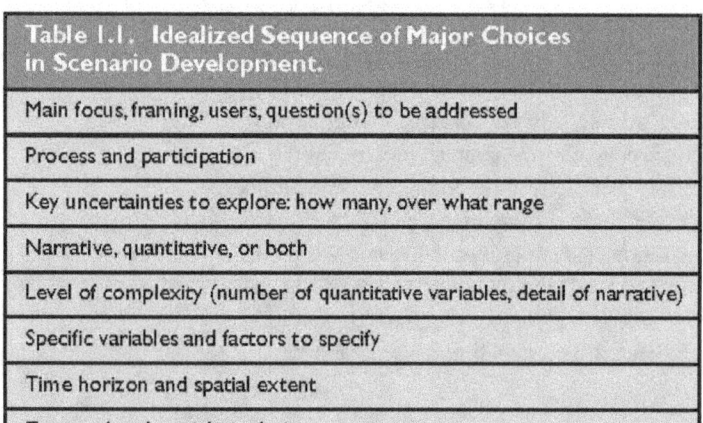

Table 1.1. Idealized Sequence of Major Choices in Scenario Development.
Main focus, framing, users, question(s) to be addressed
Process and participation
Key uncertainties to explore: how many, over what range
Narrative, quantitative, or both
Level of complexity (number of quantitative variables, detail of narrative)
Specific variables and factors to specify
Time horizon and spatial extent
Temporal and spatial resolution

---

[26] This approach does not preclude such misuse: if a goal is strongly desired, scenarios are at risk of conscious or unconscious bias to make it look easy. Japanese war-games of the Battle of Midway provide striking examples (Bracken 1977).

[27] E.g., note the predominance of scenarios on the "exploration," rather than the "decision support" side in the survey of Van Notten et al. 2003.

scenario exercises that must integrate knowledge from diverse domains, individual participants' knowledge, flexibility, and imagination can be as important as the disciplines or stakeholder groups they represent. How intensively, for how long, and by what means will participants interact? Will the process be open to outside observers or participants? How and when will feedback on the scenarios be sought, and how will it be used? How and to whom will results be communicated? And crucially, how will the be process be led, and how will disagreements be resolved? With good leadership, resolving differences in a scenario exercise can be less arbitrary and more illuminating than in other group tasks; when disagreements persist after careful examination, they can be treated as important uncertainties to be retained as alternative scenarios, not suppressed by picking a winner, splitting the difference, or retreating to vague language.

Whatever process is chosen, a series of substantive choices must be made about what goes into the scenarios. The most important of these concern what key uncertainties will be explored, and how much richness and detail should be included in scenarios to illuminate these.

What uncertainties are to be explored, and how? Many dimensions of uncertainty may be relevant to the issue being examined, but only a few can be examined explicitly in any scenario exercise. Defining these is a crucial choice that shapes much of what follows in a scenario exercise. For those uncertainties judged most important, alternative outcomes are usually represented in alternative scenarios. For example, scenarios might present high- and low-growth futures, or alternative forms that a competitive threat might take. Other uncertainties judged less crucial are usually suppressed by presenting a single "best guess" or "reference case." The few key uncertainties chosen can be represented in the number and character of scenarios, depending on the intended use. A particular uncertainty might be represented by high and low values of some quantity, or by a reference case supplemented with high and/or low variants. If two or more uncertainties interact, they can be represented by scenarios that combine different outcomes of each: in the simplest form, two interacting uncertainties can be

represented by four scenarios, often illustrated by a two-by-two matrix.[28] Several alternative scenarios might seek to span a plausible range for some key quantitative variable, or present distinct qualitative outcomes for a single uncertainty, e.g., three types of competitive threat, or three political futures for a region in turmoil. Alternatively, scenarios can represent plausible extreme or "worst-case" scenarios, to assess the robustness of decisions or strategies.

How rich and complex should each scenario be? Defining scenarios as we have, as multivariate but synoptic, still leaves a wide range of levels of complexity to choose from. At one extreme, scenarios may specify time-paths for just a few quantitative variables, or even just one. Such scenarios are common, e.g., in applications such as analyzing a firm's profitability under alternative scenarios for oil prices, or projecting tax revenues under alternative scenarios of productivity growth and inflation, often in a standard "high, middle, low" format. A scenario can accommodate more complexity by projecting additional quantitative variables, but as the number of variables increases, so also does the need for an organizing principle or gestalt to tie them together in a non-arbitrary way.

At the other extreme, the core of a set of scenarios can be a set of rich, coherent narratives, an approach frequently called the Shell approach.[29] Each narrative, described principally in text, reflects a distinct conception of how the world might develop with a persuasive underlying causal logic. A narrative scenario can stand alone but may also include specifications of important quantitative variables, e.g., of population or economic growth, consistent with the broad causal logic underlying the scenario. The narrative provides the context and explanatory logic that tie together the time-paths of quantitative variables, although particular time-paths are regarded as illustrating, not defining, the scenario – i.e., a different scenario would present substantially different time-paths or relationships among them.

Many dimensions of uncertainty may be relevant to the issue being examined, but only a few can be examined explicitly in any scenario exercise.

---

[28] Alternative interpretations of this matrix structure are discussed in van't Klooster and van Asselt 2006.

[29] Van der Heijden 1996; Wack 1985a,b; Schwartz 1991; Shell International 2003.

The choice of how rich and complex to make scenarios has far-reaching implications for the process of developing the scenarios, what can be done with them, and the uses they can serve. The two extreme approaches imply large differences in how uncertainty is treated, what aspects of the problem receive attention, and the relationship between scenarios and their users, which we discuss for climate-change scenarios in Section 4. Richer and more complex scenarios require more time and effort to develop, whether based on quantitative models, narratives, or both. Complex, narrative-based scenarios may require many person-months of development to become realistic, relevant, and persuasive, with consistent relationships among scenario elements. In return for the extra effort, this approach allows great flexibility in the way potential futures are described. Narratives can convey different aspects of a future situation with varying degrees of salience or specificity, and they can compactly convey the tone or character of a future situation by allusion, where a precise specification would appear arbitrary or labored. The narrative approach avoids limiting the defining characteristics of a scenario to any particular set of pre-specified variables, but attempts to be alert to a wide range of potentially important characteristics and mechanisms of causal influence. Proponents of this approach argue that a coherent narrative at the core of a scenario is necessary to avoid arbitrariness in specifying multiple variables, and to make the exercise useful to decision-makers: "Most scenarios merely quantify alternative outcomes of obvious uncertainties (for example, the price of oil may be $20 or $40 a barrel in 1995). Such scenarios are not helpful to decision-makers".[30]

The remaining substantive choices in specifying a scenario follow from the preceding large-scale choices. They include specifying the time horizon and spatial extent of the scenarios, deciding the particular elements to include, and the temporal and spatial resolution at which scenario outputs are stated. Decisions about temporal resolution (e.g., hourly to multi-decadal) and spatial resolution (e.g., regional, national, continental scales) are particularly important when – as is often the case in global-change applications – scenarios are produced or used by quantitative models. Such models may have very precise requirements for the specification and resolution of inputs and outputs, creating the possibility for serious mismatches between what users need or expect, and what scenario developers feel comfortable and competent providing.

This section has discussed the uses, types, and characteristics of scenarios broadly, in any application area. The next section narrows the focus to climate change and related areas of global environmental change, summarizing the types of scenarios that have been used and proposed, and that might be required, to explore and inform decision-making in this area.

---
[30] Wack 1985a: 74.

SECTION 2

Scenarios in
Global-Change Analysis
and Decision Support

Analysts have tried to develop scenarios to support understanding of and decision-making for global environmental issues, beginning with the global models of the mid-1970s and early assessment of acid rain and stratospheric ozone in the late 1970s to early 1980s.[31] The reasons for using scenarios in global change are similar to those that apply in other decision domains: high-stakes decisions that must be made under deep uncertainty about the conditions that will determine their consequences, the values at stake, or the relevant set of choices and actors. As in other domains, well-designed scenario exercises can provide a structure for assessing alternative choices and help focus on the nature of the issue, the relevant choices and actors, the values that might be at stake, and the types of research or analysis that might help clarify preferred choices.

For climate-change applications, scenario exercises have been conducted and sponsored by governments, international organizations, non-governmental organizations, and collaborative groups. While these have been diverse in form, details, and purposes, they have tended to focus more on heuristic and exploratory uses than on supporting specific decisions. The boundaries of the climate-change issue are not sharply defined, however: climate change implicates and connects to many other areas of policy, including energy, agriculture, hazard protection, and broad questions of economic development. Consequently, there is substantial uncertainty about what all the relevant decisions, decision-makers, and potentially affected values are. While some decisions are clearly of primary relevance to climate change, many other decisions that appear to be connected have not yet incorporated consideration of climate change or even recognized the connection. Reflecting these fuzzy boundaries of the issue, scenario exercises developed for climate change have overlapped with other exercises primarily focused on ecosystems, energy, and broad issues of world development. The fuzziness of the climate issue's definition increases the challenge of developing useful scenarios, but also increases the potential value of well-crafted and executed scenario exercises.

---

[31] See, e.g., Meadows et al. 1972, Barney 1981; summary of early ozone assessments in Parson 2003; and summary history of scenarios in global-change applications in Swart et al. 2004. What was the earliest scenario work in global change depends, of course, on how the boundaries of global change are defined. Kahn and Wiener 1967 might be considered an early example.

The decisions most directly related to climate change are conventionally sorted into two categories, mitigation and adaptation.[32] Mitigation consists of actions that reduce the human perturbations of the climate system, by reducing net anthropogenic greenhouse-gas emissions. Adaptation consists of actions to reduce the harm or increase the benefit from climate change and its impacts. Despite uncertainty about the precise decision agenda, we can identify in general terms the type of information scenarios might provide that would be useful to each type of decision.

Scenarios can help inform adaptation decisions by characterizing the nature and severity of relevant potential impacts; identifying key vulnerabilities, particularly those that might not otherwise have been recognized; identifying research or monitoring priorities that might give advance warning about impacts, particularly acute vulnerabilities; expanding the perceived set of potential responses; and providing a framework for evaluating alternative adaptation measures. They may also help to clarify the time structure of relevant decisions, identifying those near-term decisions that might have important but under-recognized connections to future impacts and vulnerability.

Similarly, scenarios can help inform mitigation decisions by characterizing the potential impacts of climate change and their severity, since these provide the motivation for mitigation. But, in addition, mitigation decisions can benefit from information about potential emissions trends, which determine the nature of the challenge of limiting emissions; about potential pathways of the extraction and depletion of current energy resources and development of new ones; and about potential pathways of technological development. Mitigation decisions may also benefit from scenarios representing the potential policy context in which they are made.

To date, most climate-related uses of scenarios have not examined decisions directly, but

have been embedded in larger exercises of assessment, modeling, or characterization of the issue. These uses have included formal integrated-assessment models,[33] comprehensive assessments conducted by multi-disciplinary expert bodies (e.g., IPCC), and more narrowly focused assessment exercises targeting specific aspects of the climate-change issue. In these uses, scenarios represent components of the climate-change issue that are required inputs to an assessment or model.

The causal logic of the climate-change issue is complex, including multiple two-way causal links and feedbacks among socio-economic, geophysical, and ecological systems. Integrated-assessment models seek to represent many of these linkages and feedbacks explicitly; Figure 2.1 shows a typical example of the "wiring diagrams" that illustrate the increasingly dense linkages and feedbacks represented in these models. But while such diagrams might be taken to indicate that all relationships are represented explicitly within the model – endogenously – this is not the case. All models of the climate-change issue rely on scenarios to specify some future quantities exogenously, and in virtually all cases, scenario-specified inputs are not modified to account for results of the subsequent analysis: i.e., they are truly exogenous, and the causal logic does not close.

When scenarios are used to specify exogenous inputs to a model of some aspect of the climate-change issue, the causal logic of the analysis can be greatly simplified from that shown in Figure 2.1. Instead, the logic can be represented by a simple linear structure that extends from human activities to emissions to climate change to impacts. Figure 2.2 shows this highly simplified structure. This representation is even more suitable for the uses of scenarios in other types of global-change assessments, which have been organized around much simpler causal structures than those that integrated-assessment models seek to represent. Note that we are not claiming this simple logical structure adequately represents the true structure of the climate-change issue: only that it illustrates the ways that scenarios have been used to provide exogenous inputs to global-change models and assessments.

To date, most climate-related uses of scenarios have not examined decisions directly, but have been embedded in larger exercises of assessment, modeling, or characterization of the issue.

---

[32] While this categorization has frequently been criticized for neglecting actions with overlapping effects and the third category of direct interventions in the climate system (Schelling 1983, Keith 2000, Keith et al. 2006, Parson 2006), it remains a useful approximation for most currently proposed responses.

[33] Weyant et al. 1996, Parson and Fisher-Vanden 1997.

Figure 2.1. Wiring Diagram for Integrated Assessment models of climate change. *(Source: Weyant et al 1996)*

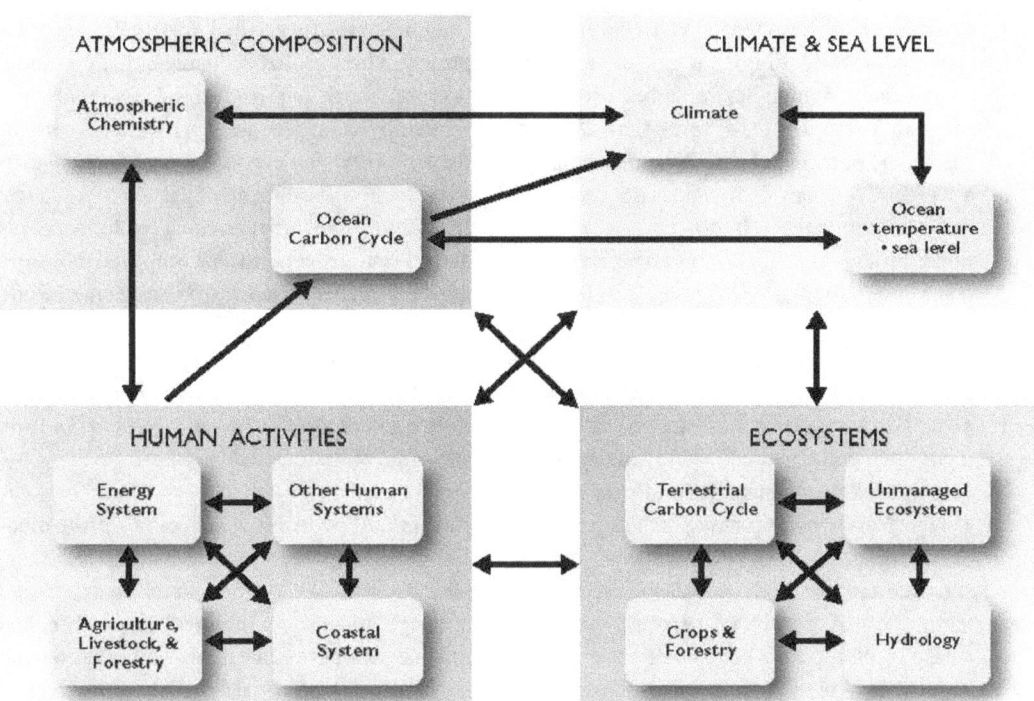

This linear logical structure allows a simple, practical categorization of five types of scenarios that have been developed for the climate-change issue. These types are defined by what quantities they specify and what primary area of analysis they provide input to. Their differences can be represented by where they cut the causal chain in Figure 2.2, with the scenario specifying quantities lying on one side of the cut, and the assessment or other activity using the scenario lying on the other side. The next five subsections discuss these five types of climate-change-related scenarios in turn.

2.1. EMISSIONS SCENARIOS FOR FUTURE CLIMATE SIMULATIONS

Scenarios of greenhouse-gas emissions, sometimes including other human perturbations such as land-use change, are the best known type of climate-change related scenario. Emissions scenarios provide required inputs to model calculations of future climate change, as shown in Figure 2.3. As the focus and intended use of climate-model studies have shifted over time, so has the role of emissions scenarios. Early research studies examined the climate system's response to potential (rather than projected) emissions inputs in individual model studies or standardized model comparisons. In such exercises, the purpose of a scenario is to provide a known, consistent perturbation big enough to generate an informative model response. Such scenarios must be standardized, so differences between model runs can be traced to scientific uncertainties and model differences, but they can be simple and arbitrary, making no claim to being realistic. The earliest such scenarios showed a "step-change" increase in atmospheric

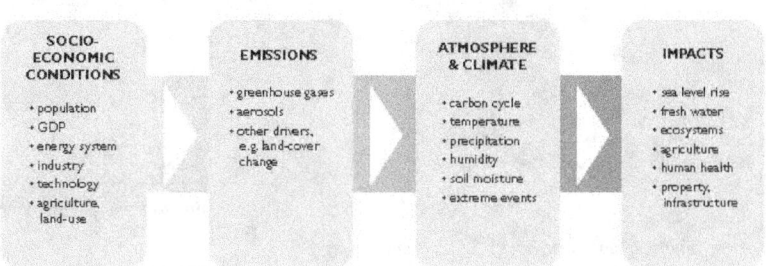

Figure 2.2. Anthropogenic climate change: Simplified linear causal chain.

concentration of CO_2 from its pre-industrial value, to either two or four times that value.[34] Models' equilibrium responses to doubled CO_2 provided a standard benchmark of model responsiveness, which has remained around the range of 1.5 to 4.5°C for more than 20 years. This range of modeled equilibrium responses to a standardized perturbation does not predict actual climate changes under human perturbations, although it has often mistakenly been taken as such.

The next generation of climate-model studies, beginning in the early 1990s, specified a time-path of atmospheric concentrations rather than a one-time perturbation. These studies for the first time allowed comparison of models' transient responses, by examining not just how much the climate changes, but how fast it changes. They still used a simple, highly idealized standard scenario, most frequently a 1 percent per year increase in atmospheric concentration, expressed as CO_2-equivalent. Only two such transient simulations had been conducted by the first IPCC assessment (1990), but by the time of the second assessment (1996), most modeling groups had produced at least one.[35]

Since the mid-1990s, climate modelers have increasingly sought to produce realistic pictures of how the climate may actually change, requiring a new approach to emissions scenarios. Scenarios must now present well-founded judgments or guesses of actual future emissions trends and their consequences for atmospheric concentrations. The required emissions scenarios have been constructed either by extrapolating recent emissions trends, or, particularly for

energy-related CO_2, representing emissions in terms of underlying drivers such as population, economic growth, and technological change and projecting these drivers using some combination of modeling and trend extrapolation. Driven by such scenarios, climate models for the first time can claim to be reasonable estimates of how the climate might actually change. In addition, comparisons using multiple models and emissions scenarios have allowed partitioning of uncertainty in future climate change into roughly equal shares attributed to uncertainty in climate science and models, and in emissions trends.[36] These comparisons have also allowed estimation of the climate-change benefits from specified emissions reductions.

As this shift occurred, advances in climate models – e.g., improved representations of atmospheric aerosols, tropospheric ozone, and atmosphere-surface interactions – produced mismatches between emissions scenarios and the input needs of climate models. For example, climate models now require emissions of several types of aerosols and reactive gases (principally the ozone precursors, hydrocarbons, CO and NO_x), explicit estimates of black carbon and organic carbon, and some disaggregation of different types of volatile organic compound (VOC) emissions. Moreover, because these emissions act locally and regionally rather than globally, they must be specified at the spatial scale of a model grid-cell, about 150 sq. km. Models of atmospheric chemistry and transport then use these emissions to generate the concentrations and radiative forcings used by the climate model. Since emissions scenarios often do not provide the required detail, climate mod-

Figure 2.3. Emissions scenarios for climate simulations.

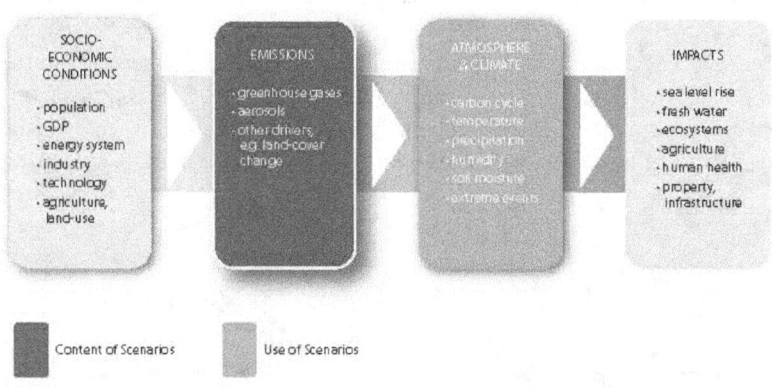

[34] e.g., Manabe and Wetherald 1967, Manabe and Stouffer 1979.

[35] Washington and Meehl 1989, Manabe et al. 1991, IPCC 1996a.

[36] Cubash et al. 2001.

elers meet these input needs through various *ad hoc* approaches.

Changes in standard emissions scenarios pose challenges for maintaining comparability with past model results. For example, the IPCC's IS92 scenarios projected that future SO_2 emissions would roughly double, then stabilize, while the later SRES scenarios projected sharp decreases, giving 2100 emissions about one-quarter the IS92 value. This change caused significant increases in projected warming that were not due to any changed scientific understanding. To help maintain backward comparability, many climate-model groups have continued to run simulations using older standardized scenarios, to provide benchmarks for comparisons both among current models and between current and previous-generation models.

2.2. EMISSIONS SCENARIOS FOR EXPLORING ALTERNATIVE ENERGY AND TECHNOLOGY FUTURES

In addition to providing needed inputs to climate models, emissions scenarios have also been produced to examine alternative socio-economic, energy, and technological futures, as shown in Figure 2.4. As in Figure 2.3 the content of the scenario is emissions, but the scenario is now used to examine the socio-economic implications of alternative emission paths, which lie upstream or to the left in the causal chain. A scenario specifying a particular emissions time-path can be used to explore what patterns of demographic and economic change, energy resource availability, and technology development are consistent with that trajectory.

Alternatively, scenarios can be used to examine what changes in policies, technologies, or other factors would be required to shift emissions from some assumed baseline onto a specified lower path, and to estimate the cost of such a shift. To be used in this way, an emissions scenario might be specified arbitrarily, or might specify some environmental target based on normative criteria as discussed in Section 1.2. Such scenarios have been most frequently used to examine emissions trajectories that stabilize atmospheric CO_2 concentrations at specified levels. More recent exercises have instead taken stabilization of radiative forcing as the target, to examine the role of non-CO_2 greenhouse gases in meeting stabilization goals.[37]

An important early example is the Wigley, Richels, Edmonds (WRE) scenarios, which presented emissions pathways that stabilized atmospheric CO_2 concentration at five levels, ranging from 450 to 1000 ppm.[38] Developed heuristically from a simple model of the global carbon cycle and two energy-economic models, these scenarios illustrated the large cost savings attainable by approaching stable concentrations through emission paths that initially rise and then decline steeply, rather than by beginning a more gradual decline immediately.

Several other sets of stabilization scenarios have been proposed and used for similar explorations. For example, the Energy Modeling Forum (EMF) has convened several multi-model scenario exercises focusing on emissions, emissions constraints, and their socio-economic effects. These have studied decision-making under uncertainty, international distribution of costs and benefits, the costs and benefits of the

Models' equilibrium responses to doubled CO_2 provided a standard benchmark of model responsiveness, which has remained around the range of 1.5 to 4.5°C for more than 20 years.

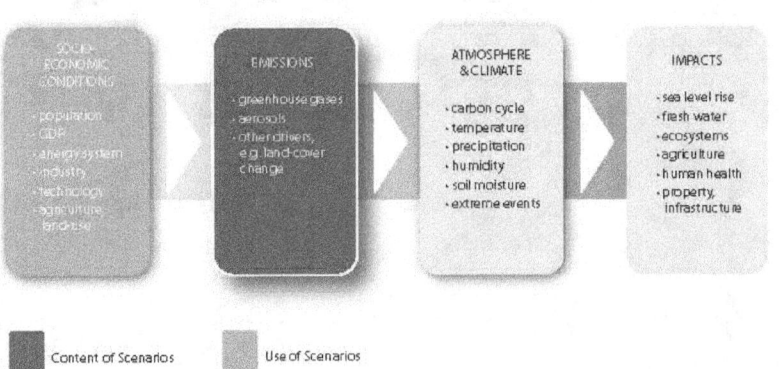

Figure 2.4. emissions scenarios for energy/technology futures.

Content of Scenarios   Use of Scenarios

[37] de la Chesnaye and Weyant 2006, EMF 2006, CCSP 2007.

[38] Wigley et al. 1997.

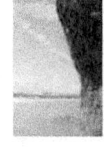

Kyoto Protocol, the implications of potential future energy technologies and technological change for emissions, and the implications of including non-CO_2 gases and carbon sequestration in mitigation targets and policies.[39]

In a recent scenario exercise of this type sponsored by the CCSP, three modeling teams constructed separate reference-case scenarios to examine the implications of stabilizing radiative forcing at levels roughly corresponding to CO_2 concentrations of 450, 550, 650, and 750 ppm. They examined the energy system, land-use, and economic implications of moving to stabilization. This project explored the role of multiple gases and alternative multi-gas control strategies in pursuing atmospheric stabilization. These scenarios may also provide a basis for future analyses by the CCSP, the Climate Change Technology Program (CCTP), or others.[40]

2.3. CLIMATE CHANGE SCENARIOS

Climate scenarios describe potential future climate conditions (Figure 2.5). They are used to provide inputs to assessments of climate-change impacts, vulnerabilities, and associated options for adaptation, and to inform decision-making related to either adaptation or mitigation. Depending on their specific use, climate scenarios may include multiple variables, such as temperature, precipitation, cloudiness, humidity, and winds. They may describe these at spatial scales ranging from the entire globe, through broad latitude bands, large continental and sub-conti-

nental regions, to climate model grid-cells or finer scales. They may project these at time resolutions ranging from annual or seasonal averages to daily or even finer-scale weather.[41]

Three major types of climate scenarios are distinguished by how they are produced: incremental scenarios, analog scenarios, and climate-model scenarios.[42] Incremental scenarios change current conditions by plausible but arbitrary amounts. For example, a region's temperature might be warmed by 2, 3, or 4°C from present conditions, or its precipitation increased or decreased by 5, 10, or 20 percent. Such adjustments can be made to annual or seasonal averages, to finer-period measurements of current conditions, or to the variability of temperature or precipitation over days, months, or years.[43] Like the simple emissions scenarios used for climate-model comparisons, incremental climate scenarios are simple to produce but make no claim to represent actual future conditions. They are used for initial exploratory studies of climate impacts and to test the sensitivity of impact models.

Analog climate scenarios represent potential future climates by the observed climate regime at another place or time. A spatial analog imposes the climate of one location on another, e.g., representing the potential climate of New York in the 2050s by that of Atlanta today or that of Illinois in the 2050s by that of Kansas today.[44] A temporal analog imposes climate conditions observed in the past, in the historical record or earlier paleoclimatic observations, e.g., using the

Figure 2.5. Climate-change scenarios.

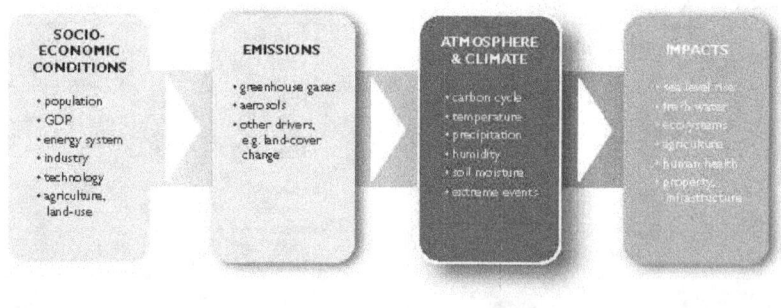

[39] See, e.g., Weyant and Hill 1999; Weyant 2004; de la Chesnaye and Weyant 2006; EMF 2006.

[40] CCSP 2007.

[41] IPCC – TGCIA 1999, Barrow et al. 2004.

[42] Mearns et al. 2001.

[43] e.g., Mearns et al. 1992, 1996; Semenov and Porter 1995.

[44] E.g., Kalkstein and Greene 1997.

hot, dry period of the 1930s to study impacts of potential future hot, dry climates.[45] Like incremental scenarios, analog climate scenarios are more useful for exploratory studies of the climate sensitivity of particular resources or systems than for projecting likely impacts. While they represent climate states that are known to be physically possible, they are limited as representations of potential future states since they do not consider the changes in greenhouse-gas concentrations that are the principal driver of climate change.

Climate-model scenarios use computers to produce a physically consistent representation of the movement of air, water, energy, and radiation through the atmosphere. Climate models, also called General Circulation Models or GCMs, approximate this calculation by dividing the atmosphere into thousands of grid-cells, roughly 150 km. square in today's models, with a dozen vertical layers, treating conditions as uniform within each cell and representing finer-scale processes by numerical relationships, called "parameterizations," that are defined at the scale of a grid cell. Climate models are used to study the present climate and its responses to past perturbations like variation in the sun's output or volcanic eruptions, and to construct scenarios of future climate change under any specified scenario of emissions and other disturbances.

Unlike incremental and analog scenarios, climate-model scenarios use emissions scenarios as inputs. Model-based scenarios have a greater claim than the other types to being realistic descriptions of how the climate might actually change, because they are based on specified assumptions of future emissions trends acting on modeled representations of known physical processes. Even with a given emissions scenario, model-based climate scenarios are uncertain. Since climate models are driven by the radiative effects of atmospheric concentrations of relevant species, some of this uncertainty comes from the carbon-cycle and chemical processes by which specified emission paths determine concentrations and radiative forcings. Some of the uncertainty can be seen in the slight differences among different runs of the same

climate model, because the models are sensitive to small differences in starting conditions. And some of the uncertainty can be seen in differences between calculations by different models, mainly caused by differences in the computational methods they use to handle errors introduced by finite grid-cells, and the parameterizations they use to represent small-scale processes.

Just as modeling future climate change requires specification of future emissions trends, assessments of future climate-change impacts require specification of future climate change. Data from a climate-change scenario might be used as input to impact assessments of freshwater systems, agriculture, forests, or any other climate-sensitive system or activity. Impact studies can use various methods, including quantitative models such as hydrologic and crop models, threshold analyses that examine qualitative disruptions in the behavior of climate-sensitive systems, or expert judgments that integrate various pieces of scientific knowledge.

As with all scenarios, the usefulness of climate scenarios depends on how well they meet users' information needs. The highly specific climate-data needs of impact analyses may not readily be provided by climate-model outputs, or may include results of whose validity climate modelers are not confident. For example, a common need of impact analyses is for data at substantially finer scales than the relative coarse grid of a climate model, which might have only 60 to 100 cells over the continental United States. One advantage of incremental and analog scenarios is that they can typically provide data at substantially finer scales. "Downscaling" techniques seek to combine the benefits of model-based scenarios – physical realism and explicit emissions-scenario drivers – and data at finer scales. The two major approaches are statistical downscaling and nested regional modeling.[46] Statistical downscaling involves estimating statistical relationships between large-scale variables of observed climate, such as regional-average temperature, and local variables such as site-specific temperature and precipitation.[47] These relationships are then assumed to remain constant under global climate

Just as modeling future climate change requires specification of future emissions trends, assessment of future climate-change impacts requires specification of future climate change.

---

[45]   E.g., Rosenberg et al. 1993.

[46]   Giorgi et al. 2001.

[47]   Wilby and Wigley 1997.

change. A nested regional climate model provides an explicit physical representation of climate for a specific region, including local factors such as mountain ranges, complex coastlines, and surface vegetation patterns, with initial and boundary conditions provided by a GCM. Regional climate models can provide projections at scales as small as 10 to 20 km. Although downscaled results are anchored to local features with well-understood climatic effects, downscaling introduces uncertainties beyond those already present in GSM results.[48]

2.4. SCENARIOS OF DIRECT BIOPHYSICAL IMPACTS: SEA LEVEL RISE

Although climate-change scenarios can provide inputs to studies of any impact, scenarios can also be constructed of particularly important forms of impact, such as sea level rise – one of the more costly and certain consequences of climate warming (Figure 2.6).[49] Changes in global mean sea level as the climate warms can be calculated using a GCM with a coupled ocean and atmosphere, which can simulate the transfer of heat to the ocean and the variation of ocean temperature with depth. To construct sea level rise scenarios for particular coastal locations, model-derived projections of global mean sea level rise must be combined with projections of local subsidence or uplift of coastal lands, as well as local tidal variations derived from historical tide-gauge data.

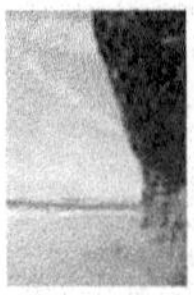

Sea level rise will increase circulation and change salinity regimes in estuaries, threaten coastal wetlands, alter shorelines through increased erosion, and increase the intensity of coastal flooding associated with normal tides and storm surge. Scenarios of sea level rise are consequently needed to assess multiple linked impacts on coastal ecosystems and settlements. In specific locations, these impacts will depend on many characteristics of coastal topography, ecosystems, and land use – e.g., coastal elevation and slope, rate of shoreline erosion or accretion, tide range, wave height, local land use and coastal protection, salinity tolerance of coastal plant communities, etc. – in addition to

local sea level rise.[50]

In addition to its gradual impacts, sea level rise is subject to large uncertainties from the potential loss of continental ice sheets in Greenland and West Antarctica. The consequences of these events for global sea level rise are well known because they can be calculated quite precisely from the volume of the ice sheets – roughly 7 meters rise from complete loss of the West Antarctic Ice Sheet and 5 meters from Greenland – but the probabilities of these events and their likely speed of occurrence are both highly uncertain. One recent study has suggested a probability of a few per cent that the West Antarctic Ice Sheet will contribute an additional one meter per century beyond that calculated from gradual warming.[51]

There are several reasons to call out sea level rise from other climate-change impacts to be represented in separate scenarios. First, sea level rise is a powerful driver of other forms of climate-change impact, probably the most important driver of impacts in coastal regions. Since it is a direct physical impact of climate change that can be described precisely and compactly, a sea level rise scenario is an efficient way to transmit the most important information about climate change to coastal impact assessments. Moreover, since sea level rise does not depend on socio-economic processes and cannot be significantly influenced by human actions (other than by limiting climate change itself), it may be reasonably treated as exogenous for purposes of impact assessment. For all these reasons, sea level rise is a good proxy for the most important causal routes by which climate change will affect coastal regions.

Finally, because sea level rise is subject to large uncertainties with known consequences but unknown probabilities, it is a useful variable for exploratory analysis of worst-case scenarios in long-range planning. Other forms of climate impact might also merit being called out in separate scenarios: changes in snowpack in mountain regions, seasonal flow regimes in major river basins, or the structure and function of major ecosystem types. Based on present

---

[48] Mearns et al. 2001, Giorgi et al. 2001.

[49] IPCC 2001a.

[50] Burkett et al. 2005.

[51] Vaughan and Spouge 2002.

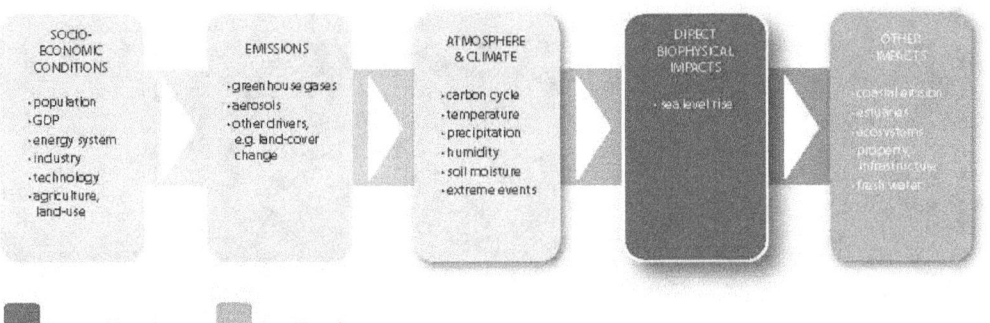

knowledge, however, only sea level rise has shown these characteristics strongly enough to motivate construction of separate scenarios.

2.5. MULTIVARIATE SCENARIOS FOR ASSESSING IMPACTS, ADAPTATION, AND VULNERABILITY

Many potentially important impacts of climate change cannot be adequately assessed by considering only how the climate might change. These impacts require multivariate scenarios that include climate change and other characteristics likely to influence impacts. This is the case, for different reasons, for both ecosystems and socio-economic systems, although the nature of the multivariate scenarios that are required – i.e., the number and identity of the characteristics that must be specified – will vary widely among particular impacts.

Ecosystems are affected by climate change, but also by many other changes in environmental conditions that are influenced by human activities, such as nitrogen and sulfur deposition, tropospheric ozone and smog, and changes in erosion, runoff, loadings of other pollutants, land use, land cover, and coastal-zone characteristics. Consequently, realistic assessments of potential future impacts on ecosystems require specifying the most important forms of human-driven stresses jointly, not just climate.[52]

In addition, many important forms of climate-change impact depend not just on climate change, its direct biophysical impacts such as sea level rise, and perhaps other forms of environmental stress, but also on the nature of the society on which these climate and other envi-

ronmental changes are imposed – e.g., how many people there are; where and how they live; how wealthy they are; how they gain their livelihoods; and what types of infrastructure, institutions, and policies they have in place.[53]

Assessment of climate impact on ecosystems that are intensively managed for human use, such as agriculture, managed forests, rangelands, and hydrologic systems, must consider human management as a factor in impacts. The non-climatic factors that influence these management decisions – e.g., changes in market conditions, technologies, or cultural practices – must be considered for inclusion in scenarios if they are sufficiently important in mediating climate impacts.

In other domains, socio-economic factors can mediate climate impacts by influencing vulnerability and adaptive capacity. No general model of the socio-economic determinants of adaptive capacity exists. Important factors are likely to vary across specific types of impact, locations, and cultures, and may include many demographic, economic, technological, institutional, and cultural characteristics.

Some socio-economic characteristics that are likely to be relevant for many impact assessments – e.g., the size and sometimes the age structure of population, the size and sometimes the sectoral mix of GDP – are normally generated in the course of producing emissions scenarios. Consequently, when current emissions scenarios exist for the region for which an impact assessment is being conducted, it makes sense to strive for consistency with them.[54] Even for these variables, however, there may be

---

52   MEA 2005.

53   Parson et al. 2001, 2003; Arnell et al. 2004.

54   Berkhout et al. 2001, citing UNEP 1994 guidelines.

**Figure 2.7:
Multivariate
scenarios for
impact assessment.**

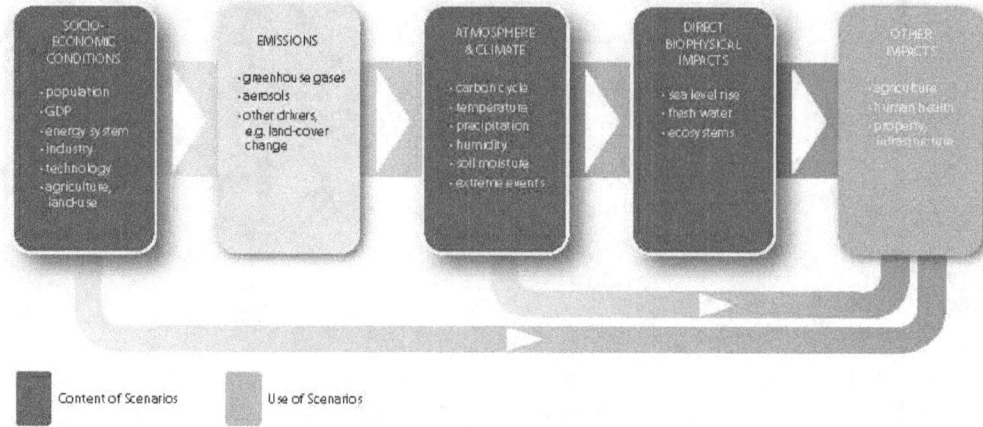

significant problems of incompatible spatial scale. Impact assessments often examine smaller spatial scales than emissions projections, so they may need these socio-economic data at finer scale than is available. Downscaling future socio-economic projections has proved challenging thus far. There is no generally accepted method for doing so, and several research groups are exploring development of alternative methods.[55]

In contrast to the few clearly identified aggregate characteristics needed to construct emissions scenarios, the socio-economic factors that most strongly shape adaptive capacity and vulnerability for particular impacts may be detailed, subtle, and location-specific. It may not even be clear what characteristics are most important before doing a comprehensive analysis of potential causal pathways shaping impacts. The most important characteristics may interact strongly with each other or with other economic or social trends, or may not be readily quantifiable. All these factors make the development of socio-economic scenarios for impact assessment a much more difficult endeavor than constructing emissions scenarios.

Because scenarios are schematic, not all factors that might be important for impacts can be included. Details are typically not included or treated as merely illustrative. But particular details, which cannot be identified in advance,

may be crucial determinants of vulnerability to climate impacts.[56] Impact assessments have responded to this dilemma in two broad ways.[57] First, constructing scenarios of relevant socio-economic conditions has been delegated to local or regional teams with expertise in the impacts being assessed, subject to constraints to maintain consistency with other assessments. Second, since local or regional scenario groups may not have access to all knowledge relevant to understanding the main determinants of impacts, more open-ended approaches have been employed – e.g., exploratory analyses that iterate between considering particular characteristics that might be important, examining their implications for impacts using the data and models available, then re-assessing what variables are most important.

This section has sketched a typology of global-change scenarios and identified major types of decision-makers who might use global-change scenario-based information. The next section turns to current experience with global-change scenarios, summarizing the development, contents, and uses of four major exercises. Informed by these cases plus additional short scenario examples presented in text boxes, Section 4 will summarize and discuss the major challenges for making and using scenarios that are raised by this experience, providing the basis for the conclusions and recommendations presented in Section 5.

---

[55] Toth and Wilbanks 2004, Pitcher 2005.

[56] Berkhout et al. 2002.

[57] Berkhout et al. 2001, Parson et al. 2001.

Review of Major Climate-Change Scenario Exercises

In this section, we review experience to date developing and using scenarios for global climate change applications. Because little literature on these activities yet exists, our selection of cases has inevitably been both limited by time and resources at our disposal and reliant in part on the knowledge and experience of team members. We discuss four exercises in detail, in an attempt to cover the largest-scale and most important activities. Section 3.1 reviews the IPCC emission scenarios, with particular detail on the most recent and important exercise, the Special Report on Emissions Scenarios (SRES). Section 3.2 considers the US National Assessment, which developed and used scenarios of both climate and socio-economic conditions. Section 3.3 considers the UK Climate Impacts Programme, which has also both developed and used scenarios, following a different approach from the US National Assessment. Section 3.4 reviews the ambitious scenario-generating exercise conducted as part of the Millennium Ecosystem Assessment (MEA), in which climate change was one of several dimensions of stress considered on global ecosystems.

For each exercise, we consider only the development and use of scenarios, rather than examining the larger assessment processes of which the scenarios were part. We consider how the scenarios were developed, including both methods of reasoning and managerial process; how and by whom they were used; and subsequent evaluations, when these are available. General issues and challenges that emerge from these experiences are discussed in Section 4.

To provide more illustrative variation, we also provide shorter summaries of eight additional scenario activities, some of them related to the four we consider in detail. Presented in text boxes throughout Section 4, these are intended to provide additional information to highlight particular issues. We have particularly sought experiences that illuminate potential relationships between scenarios and decision-making.

All these scenario exercises represent early work in an immature field. Our aim is not to criticize particular exercises, but to seek insights from their experience into the general problems of making useful global-change scenarios.

3.1. IPCC EMISSIONS SCENARIOS

Since its establishment in 1989, the IPCC has organized three exercises to develop scenarios of greenhouse-gas emissions, of increasing scale and complexity. For its first report, IPCC's Working Group 3 on "Response Strategies" included a sub-group on emissions scenarios. Four scenarios were produced but little used in this assessment because of time limits and because, with one exception, only doubled-CO_2 equilibrium climate-model runs were available at the time.[58] The next exercise produced six new scenarios, called IS92a through IS92f.[59] These were the first global emissions scenarios with a full suite of greenhouse gases and at least some explicit calculation underlying each. The IS92a scenario, one of the central scenarios in this group, was used in climate-model comparisons conducted for the 1996 IPCC assessment, along with the simpler transient scenario of 1 percent annual increase in equivalent-CO_2 concentration and further equilibrium runs.[60]

The third and most ambitious IPCC scenario process was established in 1997 and worked for two years to produce the *Special Report on Emissions Scenarios* (SRES).[61] In part, this process was established in response to two widely circulated criticisms of the IS92 scenarios. The first of these criticized the 1992 scenarios for inconsistency with other published scenarios of energy and carbon intensity for major world regions; failing to reflect important recent trends, including the collapse of the Soviet Union and increasing restrictions on sulfur emissions worldwide; relying inappropriately on a single model; and only being useful as climate-model inputs, not for other purposes such as mitigation studies or supporting climate-change negotiations.[62] The second criticized the

IS92a scenario for assuming increasing divergence in the per capita emissions of industrialized and developing regions, arguing that this represented a strong bias in favor of already developed regions.[63]

In response, the 1996 plenary session of the IPCC requested a new set of emissions scenarios. These new scenarios were to improve treatment of sulfur aerosols and emissions from land-use change, and were not to rely on a single model or expert team, but instead to draw on the existing literature and invite any group with relevant expertise to participate in an "open process."[64] They were also charged to serve more uses than climate-model inputs, such as supporting impact analyses, but to assume no new climate-policy interventions. Although not explicitly in the terms of reference, it was also clearly understood that the scenarios would address the criticism of the IS92 scenarios by focusing on convergent development paths between North and South.

In January 1997 the IPCC established a writing team, including members of several energy-economic modeling groups and experts in related areas such as population, technological change, and scenario development methods. The process ran under tight time pressure to provide preliminary scenarios by early 1998 for climate-model runs in the IPCC Third Assessment.

Prior scenarios were compiled in a web-based database,[65] and any researcher was invited to submit new ones. By mid-1998 the database contained more than 400 scenarios. Most of these projected only energy-related CO_2 emissions, but they were highly diverse in their coverage and resolution, the variables included, and their methodologies. The usefulness of these scenarios in constructing new ones was limited by several weaknesses, however. Many were incomplete, lacked documentation of inputs, or made inconsistent assumptions. Few included sulfur or land-use emissions, which were specif-

---

[58] The scenarios were mentioned in a 1-page Appendix to the Working Group 1 report. The one non-equilibrium run available was a preliminary transient run using 1 percent annual CO_2 concentration increase. See Mitchell et al. 1990, Bretherton et al. 1990, IPCC 1990.

[59] Leggett et al. 1992.

[60] The 1 percent scenario was similar to IS92a, but gave total radiative forcing about 20 percent greater by 2100. Washington and Meehl 1989; Stouffer et al. 1989; Bretherton et al. 1990:180-182.

[61] Nakicenovic and Swart 2000.

[62] Alcamo et al. 1995.

[63] Parikh 1992, 1998.

[64] Nakicenovic and Swart 2000: 324, Appendix I (terms of reference).

[65] Morita and Lee 1998.

ically requested of the new scenarios. Many were unclear on whether they assumed mitigation efforts, while the new scenarios were instructed to exclude them. Consequently, the development of new scenarios had to proceed largely independently of the collection of existing scenarios through the literature review and open process.

Early on, participants decided to use narrative scenarios in addition to quantitative models, and to include experts in this approach on the writing team. This decision drew on recent successes using such scenarios for energy and environmental applications,[66] and responded to the charge to make the scenarios more integrated and more broadly useful. Participants in an April 1997 workshop chose two key uncertainties to explore in the scenarios: whether world values would mainly stress economic prosperity or balance economic and ecological concerns (labeled "A" vs. "B" scenarios); and whether the organization of economies and institutions would continue shifting toward global integration, or reverse and move toward regional fragmentation (labeled "1" vs. "2" scenarios).[67]

Combined, these gave four scenarios, which were sketched in preliminary terms at the workshop. In the A1 (economic, global) scenario, economic growth and inter-regional income convergence continue strongly worldwide – all developing countries grow like Japan and Korea from the 1950s to the 1980s – while world population peaks at 9 billion by 2050. Rapid innovation yields many advanced energy sources, while acid rain and other local and regional environmental problems are aggressively controlled. In contrast, the A2 (economic, regional) scenario has higher population growth, lower economic growth with more continuing regional disparities, slower innovation, and weaker international institutions. B1 (ecological, global) has low population growth, moderate economic growth with strong convergence, and strong reductions in per capita energy use, mostly through higher efficiency, while B2 has intermediate population growth, low economic

growth with weaker convergence, and moderate improvements in energy efficiency and development of non-carbon energy sources.[68] The storylines were elaborated in short text descriptions with some preliminary numbers attached in fall 1997.[69]

Modeling teams were asked to produce initial quantifications of these scenarios in fall 1997, to match specified 2100 target values within 10 percent. In February 1998, the preliminary quantitative targets were re-confirmed and modelers asked to continue work on quantifications, now including a breakdown of economic output into four world regions.[70] In April, one model's quantification was chosen as a "marker scenario" for each of the four scenarios – a particular scenario that would provide the basis for interim reporting to climate modelers, some of whose results other participating models would be asked to replicate. The specifications and models for these marker scenarios are shown in Table 3.1.

These interim marker scenarios were used to provide emissions scenarios to climate models participating in the IPCC third assessment. An IPCC climate modelers' meeting in June 1998 agreed to use SRES scenarios and asked for three cases, central emissions, stabilization, and high emissions.[71] The writing team initially discussed meeting this request by identifying scenarios corresponding to each of these requested cases,[72] but decided to provide only the marker scenarios and recommend that all four be used without identifying any as "central."

The SRES interim marker scenarios were used to provide emissions scenarios to climate models participating in the IPCC third assessment.

---

66   See, e.g., WEC/IIASA 1995, WBCSD 1997.

67   Minutes, Lead Authors Meeting, Paris, April 13-15, 1997.

68   Arnell et al. 2004; Minutes, Lead Authors Meeting, Paris, April 13-15, 1997.

69   Minutes, informal modelers meeting, Berkeley, Feb 7-8.

70   Draft minutes, informal modelers meeting, Berkeley, Feb 7-8:4.

71   Minutes of the Laxenburg meeting, July 2-3, 1998, reporting results of June 29-July 1 IPCC Scoping Meeting, Bonn.

72   In July 1998, members decided that A1F or A2 could be the requested high-emissions scenario (with emissions of ~ 30 GtC in 2100), B2 or A1B a central case (~15 GtC in 2100, with two different SO2 profiles), and B1 or an A1 variant called A1R a stabilization case (at about 550 ppm) (Laxenburg report, July 2-3, 1998:1).

Table 3.1.
Target Values
for 2100 in
Initial Scenario
Quantifications

SCENARIO	A1B	A2	B1	B2
Population	7.1	15.1	7.1	10.4
GDP (trillion)	$530	$250	$340	$235
Final Energy (EJ)	~1,700	870	770	950
Model for Marker scenario	AIM	ASF	IMAGE	MESSAGE

Source: Minutes of Laxenburg meeting, July 2-3, 1998.

The SRES scenarios have been the most comprehensive, ambitious, and carefully documented emissions scenarios produced to date.

The marker scenarios also provided the basis for coordination of subsequent scenario development. Up to this point, there had been substantial discrepancy between different models' quantifications of the same scenarios, particularly at regional level. With the adoption of the markers, other groups were asked to replicate (within 5-10 percent) the marker results on population, GDP, and final energy for the four world regions, for 2100 and several interim years.[73] Achieving this requested replication posed significant challenges for modelers.[74]

With a further year of work, modeling teams produced a total of 40 scenarios that were retained in the report, of which 26 replicated one of the marker scenarios. Although a few of the 14 non-replicates reflected a model's inability to match the results of a marker scenario, most were produced because a modeling team intentionally sought to explore alternative assumptions. For example, the A1 scenario, which originally balanced fossil and non-fossil energy sources, was augmented by variants with different assumptions about fossil resources and non-fossil technology development, giving widely divergent emissions paths stressing coal, gas, and non-fossil energy technology. Modifications of the scenario set continued until late in the process. For example, it was decided in October 1998 to drop several B variants with

explicit mitigation, including one stabilization scenario.[75] At the final IPCC approval meeting, it was decided at the request of the Saudi delegation to reduce the two fossil-intensive variants of A1 to one, a variant of the gas-intensive scenario which was renamed A1FI (for "fossil-intensive").[76]

3.1.1 Significance and use

The SRES scenarios have been the most comprehensive, ambitious, and carefully documented emissions scenarios produced to date. They represented a substantial advance from prior scenarios, and contributed to assessments and subsequent research on climate impacts and responses. The SRES scenarios formed the basis for climate-model comparisons in the IPCC Third Assessment (2001) and continuing work in the Fourth Assessment. Most subsequent climate-model work has used only a few of the marker scenarios – typically A2 and B2, sometimes with A1B added. SRES scenarios also provided baselines for analysis of mitigation scenarios in the Third Assessment.[77]

Several significant insights were illuminated by the SRES scenarios.

- Scenarios with similar emissions in 2100 can follow markedly different paths in the interim, giving wide differences in cumulative emissions and concentrations.

- Technology and energy-resource assumptions can strongly perturb future emissions, even with constant socio-economic assumptions. For example, the three A1 variants show that changing these assumptions alone can generate as wide a range of emissions futures as substantial variation of demographic and economic futures.

---

73   Because markers were produced by different models with different time steps, the interim years to be harmonized differed for each scenario.

74   For example, discussions in Beijing re-confirmed that allowed deviation from markers at 4-region level would be 5 percent for GDP and 10 percent for final energy, but substantial discrepancies in base-year energy could not be harmonized due to time constraints (SRES modelers meeting report, Beijing, October 6-7, 1998:2).

75   SRES modelers meeting report, Beijing, October 6-7, 1998:4. At this meeting, B1 was also proposed for removal, but was retained based on a decision that none of the many policy interventions it presumed was an explicit greenhouse-gas limitation, so it was consistent with the terms of reference.

76   A1FI was the gas-intensive scenario, A1G, with revisions to methane emissions and additional non-CO_2 gases added from the A1 run of the MESSAGE model.

77   Morita et al. 2001.

• Highly distinct combinations of demographic, socio-economic, and energy-market conditions can produce similar emissions trajectories, suggesting that a particular emissions trajectory can pose very different types of mitigation problems, depending on what combination of driving factors underlies the emissions.

3.1.2 Criticisms and controversies

The SRES experience raised issues of great significance for subsequent attempts to develop more useful climate-change scenarios: the desirability of and appropriate methods for characterizing probabilities associated with scenarios; the quantitative representation of the relationship between North and South; methods for developing and using narrative scenarios and integrating them with quantitative model results; alternative modes for coordinating use of multiple models and their implications for the interpretation and use of scenarios; and the relationship between scenario exercises and their users, including the need for clarity about specific intended uses, appropriate methods for engaging users in scenario development, and how to improve utility of scenarios when not all potential user groups are specifically identified. These are discussed in Sections 4 and 5.

The first two of these issues were the subjects of forceful public criticisms. We discuss these, followed by several other issues that have received less attention but which in our view pose more central and instructive challenges for future scenario exercises.

Assigning explicit probabilities

The SRES team decided at the outset to make no probabilistic statements about the scenarios. Their report used great care in its language to avoid any suggestion that one scenario might be more central or more likely than any other.[78] This decision was consistent both with standard practice in developing narrative scenarios, and with the instruction in their terms of reference not to favor any model.[79]

They were sharply criticized for this decision.[80] Critics argued that there were no technical obstacles to assigning probabilities to emissions ranges bounded by the marker scenarios; that scenario developers must have made probabilistic judgments in generating and evaluating the scenario quantifications and that not making these judgments explicit would withhold relevant information; and that if scenario developers decline to assign probabilities, others who are less informed will do so. Indeed, many probabilistic emissions calculations have been produced since the SRES, using various methods such as assigning uniform or other specified distributions over the emissions range of the marker scenarios, counting scenarios lying in specified intervals in the larger SRES set, unbundling and recombining alternative values of the drivers underlying SRES emissions figures, or sampling over parameter distributions within a single model. In response to these criticisms, SRES authors argued that attempting to assign probabilities to scenarios would require assigning joint distributions to the underlying driving factors, and that this would lead to an explosion of combinatoric possibilities over which any attempt to assign probabilities would be spurious and arbitrary.[81]

The situation of the SRES scenarios is in fact more nuanced than the arguments of either their authors or critics would suggest. It may well be unhelpful to assign probabilities to rich, multi-dimensional narrative scenarios, yet still useful to assign interval probabilities when scenarios principally represent uncertainty in one or two quantitative variables. And while the SRES scenarios began their lives like the former type of storyline scenario, they finished more like the latter. For many users, the scenarios *are* their projections of greenhouse-gas emission trends. When they are viewed in this way, a potential user may reasonably ask, how likely are emissions to be higher than this – a distinct and better-posed question than, what is the probability of an A1 world? The uncertainty issue has no clear resolution in this case, and poses hard design problem for scenarios and assessments more broadly. Although the SRES exercise has

It may well be unhelpful to assign probabilities to rich, multi-dimensional narrative scenarios, yet still useful to assign interval probabilities when scenarios principally represent uncertainty in one or two quantitative variables

---

[78] E.g., Minutes of London meeting, March 1999.

[79] Draft minutes of the Washington, DC, meeting, April 29-30, 1998:6.

[80] Schneider 2001, 2002; Pittock et al. 2001; Allen et al. 2001; Reilly et al. 2001.

[81] Grubler and Nakicenovic 2001.

raised this controversy most explicitly to date, the problem is a general one that any scenario exercise must confront. We discuss it further in section 4.6.

Exchange rates: PPP versus MER

The most prominently publicized criticism of SRES focused on the fact that all but one of the participating models compared GDP across regions using market exchange rates (MER), instead of the more correct purchasing-power parity (PPP) approach. PPP comparisons correct for price differences among countries, providing a more accurate comparison of real incomes. Because lower-income countries have lower price levels, MER-based comparisons overstate the income gap between rich and poor countries.

In a series of letters to the IPCC chairman and several subsequent publications, two critics argued that the use of MER caused SRES scenarios to over-estimate future income growth in developing countries (because they over-estimated the initial income gap), and consequently to over-estimate future emissions growth. Their criticism was widely circulated and repeated by prominent climate-change skeptics.[82]

But, although using MER does overstate future income growth, it does not necessarily follow that future projections of emissions growth are also overstated. MER is universally recognized as a flawed measure of income, whose use in global-change scenarios is only justified by better availability of current and historical data, and the fact that international emissions trades in any future mitigation regime will likely be made at market exchange rates. But changing the measure of income also changes the relationship between income and such physical quantities as energy and food consumption, which determine emissions. Consequently, while MER overstates future income growth in poor countries, it also overstates future reductions in energy and emissions intensity. These opposing errors are likely to be similar in size, in which case any error in emissions projections from using MER will be small.[83]

...while MER overstates future income growth in poor countries, it also overstates future reductions in energy and emissions intensity.

A related, more serious concern is that all SRES scenarios assumed varying degrees of real income convergence between North and South; this was done in response to criticisms that the IS92 scenarios were biased in favor of the North. But an exercise to construct potential climate-change futures may need to consider less optimistic and less desirable futures in which some currently poor regions fail to solve the development problem. Not considering less fortunate futures, including ones that might challenge the adequacy of current responses, institutions, and decision-making capacity, may limit scenarios' usefulness in supporting long-term risk assessment and planning for the societal response to climate change.

Underdevelopment of narrative scenarios

Although the SRES storylines were produced first and featured prominently in publications, they remained underdeveloped and underused throughout the process. In part due to time pressure, in part due to the predominance of quantitative modelers in the process, little attention was given to further development of the storylines once initial quantifications were established and modeling work began. Nor was significant effort devoted to integration and cross-checking between storylines and quantitative scenarios, although a major purpose of the narratives was to give coherent structure to quantifications.[84] Concerns raised about the storylines included lacking specification of characteristics other than those needed to generate emissions; imbalance between the storylines, with A1 much more developed than the others and B2, the least developed, likely to be heavily used as the median scenario for emissions; apparent inconsistencies within A2; and lack of clarity regarding the distinctions between A2 and B2 – a serious enough concern that merging them was repeatedly considered until late in the process.[85]

---

[82]   Castles and Henderson 2003a, 2003b; The Economist 2003a,b; Michaels 2003b.

[83]   Nakicenovic et al. 2003, McKibben et al. 2004, Holtsmark and Alfsen 2005, Manne et al. 2005, Gr bler et al. 2004.

[84]   Minutes of the Beijing meeting, October 6-7, 1998:10.

[85]   Draft minutes of the Bilthoven meeting, September 17-19, 1997:7-8; draft minutes of the Berkeley meeting, February 7-8, 1997:6; draft minutes of the Washington, DC, meeting, April 29-30, 1998.

Moreover, participants disagreed over the meaning of some of the scenarios, as indicated by the persistent difficulty they had in agreeing on descriptive names.[86] These names were dropped late in the project, in the context of a broad retreat from attempting to flesh out the storylines. That so little integration of qualitative and quantitative components was achieved in spite of serious and persistent efforts suggests the magnitude of the analytical and methodological challenges involved.

Harmonizing scenarios, interpreting the results

Scenario exercise that use multiple models can coordinate them in several ways: choosing one or a few illustrative scenarios as coordinating devices for subsequent analyses, as was done with the SRES marker scenarios; fixing values of a small set of exogenous inputs to multiple models, to characterize resultant uncertainties and examine their origins through focused model intercomparisons; or fixing key outputs as targets, to reason backwards and examine requirements for achieving them.

Choosing a few quantitative variables as the initial link between storylines and models makes these variables serve as a framework to capture the storylines' basic logical structure. Although these choices are not obvious, the variables chosen here appear reasonable. But the causal structure of a model will not generally mirror the presumed causal logic of a narrative, so a model cannot be expected to calculate values for other variables that flesh out the storyline logic. Moreover, the few key variables so chosen may not be exogenous inputs for every model used in the subsequent quantification. Of the three variables specified in the SRES

process, only population was exogenous for all participating models. Because GDP and final energy were endogenous for some or all participating models, matching their specified values required manipulating other internal model characteristics. Once one model run was chosen as the marker for each scenario, subsequent attempts by other models to replicate the results posed the same problem more acutely, since more outputs were specified at this point.

The problems associated with attempting to harmonize model outputs are related to the underdevelopment of narrative scenarios and limited integration of qualitative and quantitative components. The storylines were associated with relatively restrictive numerical targets even though the storylines did not develop the richness or coherence that would carry implications for additional characteristics. The preliminary targets were only slightly modified throughout the project, despite subsequent discovery of significant problems. For example, the United Nations 1998 population projections, with substantial reductions in projected fertility, were completed while the scenario development work was underway but not incorporated.[87]

Clarity about uses, involving users:

The SRES scenarios were charged with serving uses beyond driving climate models but given little guidance on what specific additional uses or users to serve, or how the scenarios might best serve them, neither of which is obvious.[88] Providing climate-model inputs remained the most prominent and most clearly specified use, as well as the only use that had an early deadline. But climate modelers were not involved in the scenario development process, and there was substantial divergence between their needs and the outputs and capabilities of the SRES process. A September 1997 briefing identified the principal needs of climate modelers as early availability of scenarios and greater emissions detail.[89] They wanted separate emissions trajectories for major greenhouse gases, not just

[86] While names proposed for the "1" storylines suggest substantial common understanding (A1 was called "High Growth," "Productivity," and "Golden Economic Age," B1 was "Green" and "Sustainable development"), names proposed for the "2" scenarios, particularly B2, do not (A2 was called "Regional Consolidation," "Divided World," and "Clash of Civilizations"; B2, "Regional Stewardship," "Small Is Beautiful," "Dynamics as Usual," "Gradually Better," and "Muddling Through") (draft minutes of the Bilthoven meeting, September 17-19:7-8; draft minutes of the Berkeley meeting, February 7-8, 1997; UKCIP 1998 report summarizing SRES progress; Pitcher 1998 presentation slides.

[87] Minutes of the Bilthoven meeting, September 17-19, 1997:11.

[88] Alcamo et al. 1995.

[89] Draft minutes of the Bilthoven meeting, September 17-19, 1997:5.

CO$_2$-equivalent, including regional detail for some emissions such as sulfur – even suggesting that it would be desirable to have sulfur emissions disaggregated by stack height, to distinguish dispersed emissions from large point sources. Although SRES provided gridded sulfur data by post-processing model outputs, in most cases the emissions included and their spatial detail (not to mention stack height) were limited by the capabilities and structures of participating models.

Other uses received less attention, and representatives of other potential uses were even less involved than climate modelers in the process. Supporting assessment of mitigation strategies was largely deferred to the post-SRES scenarios prepared for the IPCC Third Assessment Report, although ambiguity about the degree of mitigation effort implied by some SRES scenarios complicated that task. Impact and vulnerability assessments depend on diverse, small-scale socio-economic and ecological factors that a global exercise centered on energy-economic models cannot provide.[90] For the population and economic projections that were provided in the course of generating emissions scenarios, the key issue for impacts and adaptation was the degree of spatial detail provided. For consistency among scenarios, and to avoid base-year discrepancies with national and regional datasets, SRES scenario results were reported only for four large world regions. Greater regional detail was available from individual models, but with inconsistent regional boundaries. Providing the greater regional detail desired for impact assessments would generate discrepancies between the global-model results represented in scenarios and the more detailed data and projections available at national and regional levels.[91] Developing valid methods to downscale socio-economic scenario information and integrate it with national and regional datasets remains a key challenge for producing useful scenarios for impact assessment.[92]

3.2. THE US NATIONAL ASSESSMENT

The US National Assessment was the most comprehensive attempt to date to assess climate impacts on the United States over 25-year and 100-year horizons, and to consider both major sub-national regions and sectors.[93] Responding to a requirement in the 1990 Global Change Research Act, the National Assessment was organized by federal agencies participating in the US Global Change Research Program. Work began in 1997, with various components completed between 1999 and 2003. Regional impacts were initially considered in 20 regional workshops, followed by more extended analyses of impacts, leading to published assessments for 12 regions, conducted by university-based teams. Sectoral impacts were examined by national teams focusing on agriculture, water, human health, coastal areas and marine resources, and forests. A federal advisory committee, the National Assessment Synthesis Team (NAST), provided direction for the assessment and synthesized its results in two published reports.[94] Roughly two thousand experts and stakeholders participated.

As an assessment focused on climate impacts, the National Assessment needed both climate scenarios and scenarios of potential future socio-economic conditions over the 21st century, since substantial changes are likely over this period in socio-economic conditions that might influence vulnerability to climate and adaptive capacity.

3.2.1. Emission and climate scenarios

For climate scenarios, the National Assessment relied predominantly on data and model results previously produced. Study teams conducted additional checking, processing, documentation, and dissemination as needed to make these

The US National Assessment was the most comprehensive attempt to date to assess climate impacts on the United States over 25-year and 100-year horizons, and to consider both major sub-national regions and sectors.

---

[90] See, e.g., discussion with Mike Hulme on behalf of TGICA, draft minutes of the Washington, DC, meeting, April 29-30, 1998:9.

[91] January 1998, meeting with Richard Moss, WG2 Technical Support Unit, described in draft meetings of the Berkeley meeting, February 7-8, 1997.

[92] Pitcher 2005.

[93] There had been two previous assessments of US climate impacts. The US EPA (1989) did a preliminary assessment for five representative US regions and five sectors (agriculture, forests, water resources, health, and coasts), while the US OTA (1993) examined impacts for six sectors – coasts, water, agriculture, wetlands, protected areas, and forests.

[94] NAST 2000, 2001.

usable. The assessment encouraged the use of three types of climate scenarios: historical scenarios produced by extrapolating observed trends or re-imposing historical climate variability or extremes; an inverse approach using sensitivity analyses to explore the responses of climate-sensitive systems, with particular emphasis on thresholds defining key vulnerabilities; and climate model simulations of future climate conditions.[95]

Of these three approaches, the climate-model scenarios were the most precisely specified and the most widely used. The National Assessment did not have the resources or time to commission new climate model runs and so had to rely on those completed and published when it began its work. A set of criteria was developed by the NAST for the climate model scenarios to be used in the assessment. Climate-model scenarios used in the assessment should, to the greatest extent possible:[96]

1. Include comprehensive representations of the atmosphere, oceans, and land surface, and key feedbacks among them

2. Simulate the climate from 1900 to 2100, based on a well-documented emissions scenario that includes greenhouse gases and aerosols

3. Have the finest practicable spatial and temporal resolution, with grid cells of less than 5° latitude x longitude

4. Include the daily cycle of solar radiation, to allow projections of daily maximum and minimum temperatures

5. Be able to represent significant aspects of climate variability such as the El Niño-Southern Oscillation (ENSO) cycle

6. Be completed in time to be quality-checked

and interpolated to the finer time and spatial scales needed for impact studies

7. Be based on well-documented models participating in the IPCC Third Assessment (for comparability between US and international efforts)

8. Be able to interface results with higher-resolution regional model studies

9. Provide a comprehensive array of results openly over the internet.

To ensure timely dissemination, the National Assessment chose climate-model scenarios to be used in its analyses in mid-1998. At that time, only two groups had completed runs that met most of the key criteria: the UK Hadley Centre (Model Version 2) and the Canadian Centre for Climate Modeling and Analysis (Model Version 1).[97] All participating regional and sector teams were asked to use these scenarios. The climate sensitivity of these models was 2.5°C (UK Hadley) and 3.6°C (Canadian), lying in the middle of the 1.7 to 4.2°C range of sensitivities represented by models participating in the IPCC Third Assessment.[98]

These two models were limited in their ability to reproduce observed patterns of inter-annual and inter-decadal climate variability. But other climate-model runs available at the time failed to meet essential requirements of the ecosystem models that were the basis for an important part of the assessment: availability of documented results, projections to 2100, standard/comparable emissions scenarios, and explicit treatment of the day-night cycle.

For these two climate models, model runs using only one emissions scenario were available, and only one ensemble run was used for each.[99] The

Of the three approaches, the climate-model scenarios were the most precisely specified and the most widely used.

---

[95] NAST 2001:25. It is arguable whether or not the inverse approach involves scenarios by the definition we have adopted here, because it does not stipulate specified future climate conditions, but attempts to identify them from presumed thresholds or breakpoints. However, we are following the usage of the NAST reports in calling these approaches three types of scenarios.

[96] NAST 2001:31-32; MacCracken et al. 2001; Mac-Cracken et al. 2003:1714.

[97] Johns et al. 1997; Boer et al. 1999a, 1999b; Mac-Cracken et al. 2003.

[98] Cubasch et al. 2001, Table 9.1:538-540; and Table 9A.1:577.

[99] Ensembles of climate-model runs are repeated simulations with small variations in initial conditions which improve the characterization of climate variability. The Canadian group had completed only one ensemble run at this time. The Hadley Center had completed three, but the National Assessment was only able to use one.

emissions scenario was IS92a, which represented the middle of the range of IPCC's 1992 scenarios.[100] In addition to greenhouse gases, the scenario included atmospheric loadings of sulfate aerosols, which were assumed to increase sharply through 2050 and then level off for the rest of the 21st century.[101]

The applicability of these two scenarios was tested by checking the models' ability to replicate broad patterns of US climate change over the 20th century when driven by historical greenhouse-gas forcings. Model results were compared against the VEMAP (Vegetation-Ecosystem Mapping and Analysis Project) dataset, a corrected climatic dataset for the 20th century. This comparison showed reasonable accuracy in reproducing the spatial distribution of average temperatures and century-long temperature trends, but significantly weaker reproduction of observed patterns of precipitation, mainly because the spatial distribution of precipitation depends on topographic detail that is too fine-scale to be captured even by the 0.5-degree VEMAP grid.[102]

With the specified scenario of future emissions, the two climate-model scenarios projected global warming by 2100 of 4.2°C (Canadian) and 2.6°C (Hadley).[103] These projections were at the high end and in the middle, respectively, of the range of warming projected for this emissions scenario by models participating in the IPCC Third Assessment Report.[104] For the continental United States, the two models projected

warming by 2100 of 5.0°C (Canadian) and 2.6°C (Hadley), at the high end and below the middle, respectively, of the range of projections in the IPCC Third Assessment.[105] In their projections of precipitation change over the United States, these scenarios both lay at the high end – the Hadley scenario projected the highest precipitation in 2100 and the Canadian the second-highest[106] – but the Canadian model's greater warming offset the effect of this precipitation increase on soil moisture, which was projected to decrease over most of the continental United States.[107]

Although only the Hadley and Canadian climate-model scenarios were used throughout the assessment, several others that met some or all of the assessment's needs became available during its work. Several region and sector teams were able to use these additional scenarios. In some cases, the additional scenarios allowed groups to strengthen their conclusions. For example, an analysis of future Great Lakes water levels under climate change using eleven climate models found that ten of these showed lower levels and only one higher.[108] In other cases, using multiple models allowed more detailed characterization of uncertainties in future regional changes. For example, the Pacific Northwest team presented distributions of regional temperature and precipitation change in the 2030s and 2090s using seven GCMs.[109]

Despite the National Assessment's aim of exploring future climate using three distinct types of scenario, historical scenarios and sensitivity analyses were less extensively used than GCM scenarios and featured less prominently in the

[100] The IS92a scenario is described in section 3.1. There were small differences among climate-modeling groups in the way they converted emissions trajectories into atmospheric concentrations and radiative forcings, making the actual scenarios driving each model run very close, but not quite identical.

[101] See www.usgcrp.gov/usgcrp/nacc/background/scenarios/emissions.html for further detail on emissions scenarios used in the National Assessment.

[102] VEMAP members 1995, Kittel et al. 1995.

[103] NAST 2001:36, Table 2.

[104] Cubasch et al. 2001, Figure 9.5a:541. While the Canadian model lies at the high end, it is not an outlier. The GFDL model (which was more responsive than the Canadian model, with a climate sensitivity of 4.2° C) projected higher global warming than the Canadian model in this scenario for the first few decades of the century, but only had results through 2060 in time for the TAR.

[105] The seven models for which these results were available clustered at the top and the bottom. Three of them – the Canadian, GFDL, and Hadley 3 models – lay very close together at the high end, the Canadian the highest by a fraction of a degree; three others lay close together at the low end, Hadley 2 the highest of them by somewhat less than a degree. A seventh model, ECHAM4, tracked the high group through 2050, the last year for which its results were available. Since these comparisons usually reflect only one ensemble run of each model, small differences between runs may reflect consistent inter-model differences, or noise reflected in a single ensemble run. See NAST 2001:547, Figure 7.

[106] NAST 2001:545, Figure 8.

[107] NAST 2001:552, Figure 16 and 18.

[108] Lofgren et al. 2000; NAST 2001:175.

[109] NAST 2001:256.

assessment's publications. Two uses of historical climate data – describing observed impacts of climate variability and using observed historical extremes as benchmarks to compare projected future changes – were made by all groups. To support systematic use of historical scenarios, the VEMAP 20th-century dataset was provided to all groups, but no further guidance was provided on how to generate climate scenarios from these historical data, e.g., on what periods to choose or how to use them to assess potential future impacts. Several groups used these historical data to describe the impacts of particular recognized patterns of climate variability, such as ENSO or the Pacific Decadal Oscillation (PDO).[110] Many groups examined past climate extremes, but only in qualitative ways; most did not follow the approach, taken in some previous impact studies, of using historical extreme periods as quantitative proxies for potential future climate.[111]

The third approach, vulnerability analysis, was the least used in the National Assessment. This "inverse" approach involves describing the properties of a climate-sensitive system, specifying some important change or disruption, and asking what climate changes would be required to bring about that disruption and how likely – based on historical data and model calculations – such climate changes appear to be. Given the complex dynamics of climate-sensitive systems and models of these systems, and the multiple dimensions of climate on which these can depend, this approach requires a substantial program of new research, analysis, and methodological development.[112] In part because of the intrinsic difficulty of this task – and in part due to management and resource problems – this approach was not pursued. The NAST proposed it, but more tractable approaches to analyzing climate impacts dominated the assessment's work. This remains an important area for further work in development of assessment and modeling methods.

---

[110] E.g., Mote et al. 2003, Southeast Regional Assessment Team 2002.

[111] Rosenberg et al. 1993.

[112] For an example of such efforts, see the AIACC (Assessments of Impacts and Adaptations to Climate Change) project, information at http://www.aiaccproject.org.

3.2.2. Socio-economic scenarios

As discussed in Section 2.5 above, assessing impacts of future climate change can require specifying not just scenarios of future climate, but also socio-economic characteristics of the future society that will experience the changed climate. Specifying future socio-economic conditions might be necessary for two reasons. First, socio-economic conditions may influence the demands placed on particular resources that are also sensitive to climate change, the value assigned to them, and the non-climatic stresses imposed on them. For example, future flow regimes in river systems will be influenced by upstream demands for municipal and irrigation water use, in addition to the changes caused by climate. Socio-economic scenarios are also needed to assess climate-change impacts on human communities – e.g., economic impacts and their distribution, human health effects, and vulnerability to extreme events – because socio-economic characteristics of a community experiencing a changed climate will strongly influence the community's vulnerability to changes and its capacity to adapt.

In contrast to climate scenarios, little prior information or experience was available on constructing scenarios of socio-economic conditions for impact assessment. Consequently, the assessment developed new methods, using an approach that combined centralized and decentralized elements. Centralization was needed because a few variables, such as population, economic growth, and employment, are likely to be important in all regions and sectors. For these, consistent assumptions are required to allow comparison of impacts across regions and sectors, and to aggregate from separate assessments up to overall national impacts. A NAST sub-group developed high, medium, and low-growth scenarios of these variables at the national level. These followed the US Census Bureau high, middle, and low scenarios for fertility and mortality through 2030, but assumed a wider range of values for net immigration to account for possible illegal immigration.[113] Over this period, national population, GDP, and employment were disaggregated among regions and sectors using a commercial regional eco-

---

[113] Parson et al. 2001:102-103.

Socio-economic scenarios are also needed to assess climate-change impacts on human communities.

nomic model.[114] Beyond 2030, national projections of these variables followed OECD growth rates in the SRES marker scenarios.[115]

Decentralization was also needed because the particular socio-economic characteristics that most strongly influence climate impacts and vulnerability may differ among regions, activities, and resources. For example, major socio-economic determinants of climate impacts on Great Plains agriculture may include the crops grown, the extent of irrigation, and the technologies used to provide it, while the main determinants of coastal-zone impacts may be patterns of coastal development, zoning, infrastructure, and local property values. The NAST judged that those assessing regional or sector impacts were likely to know more about such factors than a central body. Consequently, to support decentralized scenario development, the NAST proposed a consistent template for assessment teams to follow in creating their own scenarios. Teams were asked to identify two socio-economic factors they judged most important for their impacts of concern; to identify a range of these factors to represent roughly 90 percent confidence; and to create socio-economic scenarios by combining high and low values of these factors, plus middle or best-guess values if they so chose.

Implementation of socio-economic scenarios in the National Assessment was weak. Few assessment teams used the proposed approach. Many made no socio-economic projections at all, but rather projected only biophysical impacts based on GCM results. One assessment team found the socio-economic scenarios were inconsistent with superior local estimates of current population, and so decided not to use them.[116] The teams that did use the socio-economic scenarios used only aggregate projections of population and economic growth, or in some cases assumed continuation of present conditions in the assessment period. None used the proposed template for identifying and projecting additional important socio-economic characteristics.

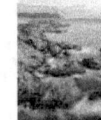

More useful assessments of impacts and vulnerability will require more extensive use of socio-economic scenarios, improved integration of socio-economic with climatic and environmental scenarios, and substantial further investment in development and testing of new methods.

Several factors contributed to this limited use of socio-economic scenarios. In addition to various managerial and communication problems, many participants were reluctant to use socio-economic scenarios, especially the proposed decentralized approach. Some preferred to avoid any socio-economic projections, implicitly presuming either that socio-economic conditions did not matter for impacts, or that those that did matter would remain similar to present conditions. Others objected to specific contents of the scenarios or the methods used to generate them, or judged that their team lacked the expertise required to evaluate them. Still others objected that uncertainties in future socio-economic conditions made any attempt to construct scenarios for more than a few years in the future unacceptably speculative.[117] Consequently, while the assessment attempted to advance scenario methods, weak implementation of these methods limited its ability to identify key vulnerabilities. More useful assessments of impacts and vulnerability will require more extensive use of socio-economic scenarios, improved integration of socio-economic with climatic and environmental scenarios, and substantial further investment in development and testing of new methods.[118]

3.2.3. Criticisms and controversies

The National Assessment was the object of substantial political and scientific controversy. Here, we summarize the major criticisms that pertain to the development and use of scenarios. Criticisms focused predominantly on the climate scenarios, especially those derived from GCMs, probably because these were more precisely defined, widely used in the analyses, and featured in the assessment's publications. Three criticisms of these were advanced.

The first, widely circulated during 2000, was that the use of non-American climate models for climate scenarios was inappropriate and potentially injurious to national interests.[119] While this criticism indicates a dimension of political

---

[114] Terleckyj 1999a,b.

[115] The high-growth scenario was roughly comparable with A1, medium with B1, and low with A2 and B2.

[116] Rosenzweig and Solecki 2001.

[117] Morgan et al. 2005.

[118] Lorenzoni et al. 2000, Berkhout and Hertin 2000, Parson et al. 2003.

[119] Congressional Record, June 16, 2001, Statements of Senators Hagel (page S5292) and Craig (page S5294).

vulnerability of the assessment, it does not address the assessment's technical quality. Since climate models represent the physics of the global atmosphere, they contain no representation of political or economic factors. The Hadley and Canadian global climate models were extensively documented in peer-reviewed scientific literature – and, moreover, were the only models that met the most critical of the assessment's criteria. That they were developed by scientific groups outside the United States has no significance for their ability to provide scenarios to assess US impacts. Using US models would have avoided this criticism, but at the cost of either weakening the analysis by using scenarios that did not meet the assessment's needs, or delaying the work by one to two years. In deciding to proceed with non-US models, assessment organizers judged that these costs were too high.

The second major criticism was that the two climate-model scenarios used were at the extreme end of available models in their projected climate change. This is partially correct. When temperature and precipitation factors are considered together (because high precipitation in some cases may offset the impacts of high temperature), the Canadian scenario lies at the high-impact end – although not an outlier, as other IPCC model projections lie close to it – while the Hadley lies at or somewhat below the middle for most analyses.

The National Assessment's organizers and its critics agreed that using more models would have been preferable, but the assessment was limited by its schedule and its technical requirements. Given a limit of only two, there can be good reasons to choose one scenario in the middle of current projections and one near the top that provides a plausible upper-bound, but the significance of the results must then be communicated with great care. Some critics suggested that presentation of results based on the relatively high Canadian scenario should be more carefully qualified to highlight its position near the top of current projections.[120] Such qualifications must be crafted very subtly, however, lest they imply these results may safely be ignored, when most analyses suggest the full

range of future climate-change uncertainty extends both below the Hadley scenario and – in a long, thin tail – above the Canadian.

A related criticism of the climate scenarios claimed that the emissions scenario driving them was implausibly high. The issues bearing on choice of emission scenarios are similar to those for choice of climate models. It would be preferable to have a wide and relevant range of emissions scenarios driving an impact assessment – at least for the post-2050 period. Using a wide range of emissions scenarios would also allow comparison of projected impacts under high and low emissions futures, and so give insights into what degree of impacts could be avoided by what degree of mitigation effort. Model runs with this emissions scenario were all that were available, however. Moreover, there is no clear basis to reject this particular scenario, since it was the scenario most widely used in climate-model runs at the time and lies near the middle of the range of both the 1992 and the 2001 IPCC scenarios. Finally, there is no support for the claim that this scenario was chosen with the aim of making 21st-century climate change appear as frightening as possible.[121] But, although using just two climate models with one emissions scenario was unavoidable in this assessment, it still represented a serious limitation. With more model simulations testing a range of emission scenarios already available, future assessments will be able to remedy this deficiency.

In contrast with the preceding criticisms that the scenarios used in the assessment understated uncertainty, another criticism focused on the disparities between the two scenarios' projections. Some critics argued that such disparities – e.g., the Canadian scenario projects the Southeastern states becoming much drier than the Hadley model does – show that our limited knowledge of regional climate change makes any attempt to assess future impacts and vulnerabilities irresponsible.[122] This criticism im-

---

[120] MIT Integrated Assessment project, comments on National Assessment, Aug 11, 2000:15.

[121] Michaels 2003a:171-192.

[122] Disparities between the two models' projections were the basis of an unsuccessful lawsuit brought against the Assessment under the Federal Data Quality Act (See Competitive Enterprise Institute, "Complaint for Declarative Relief," http://www.cei.org/pdf/3595.pdf, at paragraph 24.)

plies that impact assessment should wait until precise, high-confidence regional climate projections are available. Since a major purpose of the assessment was to represent current uncertainty about climate change and its impacts, such discrepancies between model projections served a valuable purpose, as indications of the uncertainty of projections at the regional scale – particularly when the model disparities had a clear origin, such as differences in projected jet-stream location.

In sum, the National Assessment's use of climate-change scenarios was hampered by the lack of available relevant runs, but reflected an adequate attempt to represent then-understood variation in climate projections for the United States. Future assessments will need to use more climate-model projections – including multiple ensemble runs – informed by a wider range of relevant emissions scenarios. The National Assessment attempted to advance the state of the art in using socio-economic scenarios, but achieved only limited success in implementing its plans. Future assessments will need to invest substantial resources in developing the state of underlying knowledge, models, and assessment methods for integrating socio-economic considerations into assessments of climate impacts. This includes further development of novel approaches to link climate and socio-economic scenarios, such as the proposed "inverse" approach to vulnerability analysis.

The experience of the National Assessment raises three significant issues for future climate-change scenario exercises. First, like several of the experiences reviewed here, it illustrates the difficulty and scale of effort involved in producing scenario-based assessments. Second, the large required start-up effort and time to build the capacity to conduct such an exercise illustrates the great value of sustaining analytic and institutional capacity over time, rather than relying on separate projects. Such continuity of capacity will avoid wasteful repetition of start-up efforts, support accumulation of learning and experience, and develop and maintain the required collaborative networks. Finally, the assessment's experience illustrates both the need for consistency in large-scale assessments, and the great specificity of information needs within particular impact and adaptation assessments.

This combination of centralized and decentralized information requirements suggests the need for a cross-scale organizational structure for developing and applying scenarios, including scenarios of both climate and socio-economic conditions.

3.3. THE UK CLIMATE IMPACTS PROGRAMME

The UK Climate Impacts Programme (UKCIP) was established in April 1997 as one element of a broad program of scientific research, assessment, and support for policy-making on climate change. The UKCIP supports research and analysis of impacts for particular regions, sectors, and activities in the UK. The program provides common datasets and tools, as well as ongoing support to university researchers and organized stakeholder groups in all UK regions. As part of its role in stimulating, supporting, and coordinating decentralized and stakeholder-driven impact analyses, the UKCIP has produced and disseminated three sets of scenarios: climate scenarios in 1998 and 2002, and socio-economic scenarios in 2001.

The 1998 climate scenarios provided information only at the rather coarse scale of the Hadley Centre's HadCM2 climate model, with four grid-cells over the entire UK. Four scenarios, called "high," "medium-high," "medium-low," and "low," combined variation in emissions assumptions with variation in assumed climate sensitivity. The medium-high and medium-low scenarios both used the HadCM2 model, with a sensitivity of 2.5°C.[123] The medium-high scenario was driven by a 1 percent per year equivalent-CO_2 transient scenario, similar to IS92a. The medium-low scenario was driven by a 0.5 percent per year equivalent-CO_2 transient scenario, similar to the lowest IS92 scenario, IS92d. The high and low scenarios used the same two emissions scenarios driving a simpler climate model, whose sensitivity was set at 4.5°C for the high scenario and 1.5°C for the low. These scenarios were used in an initial impact assessment focusing predominantly on direct biophysical impacts.[124] The scenarios did not include any explicit statements of probabil-

Future assessments will need to invest substantial resources in developing the state of underlying knowledge, models, and assessment methods for integrating socio-economic considerations into assessments of climate impacts.

---

[123] UKCIP 1998:13-15.

[124] UKCIP 2000.

ity, although their documentation suggested that the medium-high and medium-low scenarios "in one sense ... may be seen as being equally likely,"[125] while the high and low scenarios captured part of the tails of the distribution. Nor did they include any potential extreme climate events such as those associated with large changes in the North Atlantic circulation.

The UKCIP's socio-economic scenarios drew on the Foresight Program, a broader exercise of the UK Department of Trade and Industry to develop scenarios for long-range planning in several policy areas, with additional detail in areas relevant to greenhouse-gas emissions and climate impacts.[126] As in several other scenario exercises, developers identified two fundamental uncertainties and combined two alternative outcomes of each to produce four scenarios. The two core uncertainties they chose were similar to those used in the SRES exercise: social and political values, which varied from an increased focus on individual consumption and personal freedom ("consumerism") to a widespread elevation of concern for the common good ("community"); and governance, which varied from authority and power concentrated at the national level ("autonomy"), to power increasingly flowing to global institutions, downward to local ones, and outward to nongovernmental institutions and civil society ("interdependence"). The two dimensions of uncertainty, values and governance, were assumed to be independent of each other. Other major uncertainties such as demographic change, the rate and composition of economic growth, and the rate and direction of technological change, were treated largely as consequences of alternative realizations of the two core dimensions of values and governance.[127]

The four scenarios built around these two dimensions of variation were called "National Enterprise," "World Markets," "Local Stewardship," and "Global Sustainability." Each was initially developed as a qualitative narrative of future conditions in UK society intended to apply broadly to both the 2020s and 2050s. Each scenario specified several dozen socio-

economic characteristics qualitatively, including multiple aspects of economic development, settlement and planning, values and policy, agriculture, water, biodiversity, coastal zone development, and the built environment.[128]

Each scenario was also realized in projections of multiple quantitative variables, at the national scale only. For the 2020s, these provided detail on population, GDP (including the governmental share and the sector split between industry, agriculture, and services); household numbers and average household size; land use and rates of change; total transport and modal split; agricultural production (including such details as chemical and financial inputs, subsidies, yields, and organic area); freshwater supply, demand, and quality; and several indicators of biodiversity and coastal vulnerability. For the 2050s a smaller set of quantitative variables was projected, describing population, GDP, land use, and transport. The plausibility of projections was checked, mainly by comparing projected future rates of change to historical experience. The scenarios were published with a detailed guidance document, which provided suggestions on how to use them together with climate scenarios for impact studies.[129]

As of 2005, the socio-economic scenarios had been used in six impact studies.[130] There has been some difficulty applying the national-level scenarios in specific, smaller-scale regions. The most ambitious use has been a preliminary integrated assessment of climate impacts and responses in two regions of England, the Northwest and East Anglia.[131] This study produced four integrated scenarios of regional climate impacts, by pairing each of the four socio-economic scenarios with one climate scenario based on a rough correspondence between the socio-economic scenario and the IPCC emissions scenario underlying the climate scenario[132] Based on these four scenarios, the

The UKCIP's socio-economic scenarios drew on the Foresight Program, a broader exercise of the UK Department of Trade and Industry to develop scenarios for long-range planning in several policy areas, with additional detail in areas relevant to greenhouse-gas emissions.

---

[125] UKCIP 1998:iv.

[126] UKCIP 2001.

[127] UKCIP, 2001.

[128] Berkhout et al. 2001.

[129] Berkhout and Hertin 2001.

[130] UKCIP 2005.

[131] Holman et al. 2002.

[132] Regional (National) Enterprise was taken as UKCIP High (IPCC A2); Global Markets as UKCIP Medium-High (A1B); Regional (Local) Stewardship UKCIP Medium-Low (B2); and Global Sustainability UKCIP Low (B1).

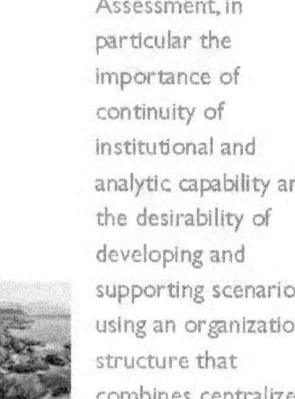

The UK program's experience highlights some of the same issues for future scenario exercises as the US National Assessment, in particular the importance of continuity of institutional and analytic capability and the desirability of developing and supporting scenarios using an organizational structure that combines centralized and decentralized elements.

study elaborated preliminary regional scenarios corresponding to the four national socio-economic scenarios, and conducted an assessment of coastal-zone impacts and responses using these scenarios and a formal land-use model.

Four new climate scenarios were produced in 2002, based on the SRES marker scenarios and new versions of Hadley Center climate models. These new scenarios differed only in their emissions assumptions, not climate sensitivity. The high, medium-high, medium-low, and low scenarios were driven by the A1FI, A2, B2, and B1 marker scenarios, respectively. These were used to drive the HadCM3 global climate model (with a grid-scale of 250-300 km), generating climate-change projections for 30-year future periods centered on the decades of the 2020s, 2050s, and 2080s. For some emissions scenarios and time periods, climate projections were processed through a nested hierarchy of three Hadley Center climate models: the HadCM3 model at global scale, the HadAM3H model at intermediate scale, with a horizontal resolution of about 120 km, and the HadRM3 model for high-resolution climate projections in the United Kingdom and Europe, with a horizontal resolution of about 50 km. This nested processing was done for the baseline period (1960-1990), and for the most distant projection period (2070-2100) to produce three ensemble runs for the medium-high (A2) emissions scenario and one for the medium-low (B2). For the other emissions scenarios and the intervening projection periods, results of the global-scale model were downscaled using statistical patterns of fine spatial-scale climate variation derived from full runs using scenario A2. These scenarios were widely distributed and supported through a web-based interface, including map-based graphical display of projected changes in more than a dozen climate indicators on a fine-scale (50 km) grid of the United Kingdom.

Several analyses are continuing to use the 2002 climate scenarios in conjunction with the socio-economic scenarios. For example, a 2004 integrated analysis of flood risk and erosion control over a 30-100 year time horizon produced a

threat assessment, a set of scenarios of flood risk, and a set of policy recommendations. An evaluation of this study's effects one year later found that it was being used by several public and private actors to inform decision-making.[133]

The UKCIP, in contrast to the US National Assessment, has built a sustained assessment capability. In addition, the central program has less authority over the separate assessments, instead acting more as motivator, resource, and light coordinator. Access to scenarios is to licensed users, of whom there are about 130 – roughly half in universities, the rest about equally split among private sector and all levels of government. Most active users have been national officials responsible for climate-sensitive resources, with less participation from the private sector and local governments.[134]

The program has invested in generating, disseminating, and documenting useful climate scenarios for impacts users. The jury appears to still be out on whether the level of effort and success is similar for socio-economic scenarios, which have not yet been either downscaled or repeated. Getting scenarios used is a slow process, but the scenarios produced by this program are starting to be used by decision-makers in support of their practical responsibilities. A significant limitation of the program, however, is its exclusive reliance on just one family of climate models. This may pose risks of under-estimating future climate uncertainty and over-confidence in assessments of potential climate impacts and responses. Although the UK program followed a substantially different organizational model from the US National Assessment, its experience highlights some of the same issues for future scenario exercises, in particular the importance of continuity of institutional and analytic capacity and the desirability of developing and supporting scenarios using an organizational structure that combines centralized and decentralized elements.

[133] UK Office of Science and Technology 2002.

[134] West and Gawith 2005.

3.4. THE MILLENNIUM ECOSYSTEM ASSESSMENT

The Millennium Ecosystem Assessment (MEA) was a large, United Nations (UN)-sponsored assessment of the current status, present trends, and longer-term challenges to the world's ecosystems, including climate change and other sources of stress. Conducted between 2001 and 2005, the MEA sought to assess changes in ecosystems in terms of the services they provide to people and the effects of ecosystem change on human well-being. It also sought to identify and assess methods to mitigate and respond to ecosystem change, for various private and public-sector decision-makers, including those responsible for the several international treaties that deal with ecosystems.[135] More than 1350 authors from 95 countries participated in the global assessment's four working groups, and hundreds more in about 30 associated sub-global assessments. The assessment's goals were broad, ranging from providing a benchmark for future assessments and guiding future research to identifying priorities for action.[136]

Results of the global assessment were presented in a March 2005 synthesis report, and in additional volumes presenting the output of the assessment's four working groups, "Current State and Trends," "Scenarios," "Policy Responses," and "Multi-Scale Assessments." The current state and trends group examined ecosystem trends over the past 50 years and projections to 2015; the scenarios group took a longer view to 2050 and beyond. Because of time limitations, the work of these two groups proceeded largely independently.

All components of the assessment used a common large-scale conceptual framework, which distinguished indirect drivers of ecosystem change, direct drivers, ecosystem indicators, ecosystem services, measures of human well-being, and response options. Direct drivers included direct human perturbations of the environment such as climate change, air pollution, land-use and land-cover change, resource consumption, and external inputs to ecosystems such as irrigation and synthetic fertilizer use. Indirect drivers included underlying socio-economic factors such as population, economic growth, technological change, policies, attitudes, and lifestyles.[137]

The scenarios working group sought to apply this conceptual framework to long-term trends in ecosystems, looking ahead to 2050 with more limited projections to 2100. They developed the structure of the scenarios in an iterative process, including consultations with potential scenario users and experts in a wide range of decision-making positions around the world.[138] Like several other major scenario exercises, they initially sought to identify two basic dimensions of uncertainty in long-term ecosystem stresses, which together would produce four scenarios.[139] For the first dimension, like SRES they chose globalization: continuation and acceleration of present global integration trends, versus reversal of these trends to increasing separation and isolation of nations and regions. For the second dimension, in contrast to the broad value-based uncertainties used in the SRES and UKCIP scenarios, they chose one more specifically related to ecosystems: whether responses to increasing ecosystem stresses are predominantly reactive – waiting until evidence of deterioration and loss of services is clear – or predominantly proactive, taking protective measures in advance of their clear need. The combination of two polar values of each of these uncertainties yielded four scenarios, summarized in Table 3.2.

The Global Orchestration (global, reactive) scenario presented a globally integrated world with low population growth, high economic growth, and strong efforts to reduce poverty and invest in public goods such as education. In this scenario, society focuses on liberal economic values, follows an energy-intensive lifestyle with no explicit greenhouse-gas mitigation policy,

The MEA sought to assess changes in ecosystems in terms of the services they provide to people and the effects of ecosystem change on human well-being.

---

[135] E.g., the Convention on Biological Diversity, the Convention to Combat Desertification, the Convention on Migratory Species, and the Ramsar Convention on Wetlands.

[136] MEA 2006:xii.

[137] MEA 2006:153 (Table 6.1) and 304 (Table 9.2).

[138] MEA 2006:152.

[139] MEA 2006, Figure 5.2.

**Table 3.2.
Millennium
Ecosystem
Assessment
Scenarios**

ECOSYSTEM MANAGEMENT	WORLD DEVELOPMENT	
	Global	Regional
Reactive	Global Orchestration	Order from Strength
Proactive	TechnoGarden	Adapting Mosaic

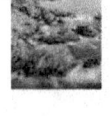

and takes a reactive approach to ecosystem problems.[140] In Order from Strength (regional, reactive) the reactive approach to ecosystem problems takes place in a fragmented world pre-occupied with security and less attentive to public goods.[141] This scenario exhibits the highest population growth and lowest economic growth. Economic growth is particularly low in the developing countries, and it decreases over time. In Adapting Mosaic (regional, proactive), political and economic activity are concentrated at the regional ecosystem scale. Societies invest heavily in protection and management of ecosystems in locally organized and diverse efforts. Population growth is nearly as high as in Order from Strength, and economic growth is initially slow but increases after 2020. Finally, TechnoGarden (global, proactive) presents a world that is both focused on ecosystem management and globally connected, with strong development of environmentally friendly technology. Population growth is moderate, and economic growth is relatively high and increasing.[142]

Each scenario was initially constructed as a qualitative description. Population and GDP were specified quantitatively, while all other indirect drivers – including social, political, and cultural factors – were qualitative. Population scenarios were derived from the International Institute for Applied Systems Analysis' (IIASA's) 2001 probabilistic projections, capturing the middle 50-60 percent of the distribu-

tion, with world population in 2050 ranging from 8.1 billion (Global Orchestration) to 9.6 billion (Order from Strength).[143] No statements of probability or likelihood were made about the scenarios.

From the indirect drivers, a more specific and quantified set of direct drivers was developed, using formal models where possible. Species introduction and removal was the only unquantified direct driver.[144] Separate pre-existing models were used of the world energy-economy, greenhouse-gas emissions and climate change, air pollution, land-use change, freshwater, terrestrial ecosystems, biodiversity, and marine and freshwater fisheries. To the extent possible, these quantitative models were used to reason from indirect and direct drivers to ecosystem effects, changes in ecosystem services, and effects on human well-being.[145] In some cases this was achieved by soft-linking models, using outputs from one as inputs to another, but this was limited by different variable definitions, spatial and temporal resolution, and other model incompatibilities.[146] Not all scenario elements could be modeled quantitatively, so expert judgments were also extensively used. The qualitative scenario process proceeded in parallel with quantitative modeling – elaborating aspects of the scenarios that were not amenable to modeling, filling gaps, and stipulating feedbacks between ecosystem services and human well-being and behavior.[147]

---

[140] MEA 2006, Ch 5.5.1

[141] This scenario was originally named "Fortress World" (report of first meeting of MEA global modeling group, Jan 7, 2003). The later name reflected participants' judgments that in such a decentralized world preoccupied with security concerns, maintaining global order would require democratic nations to be militarily strong – i.e., it is a world of "realist" international affairs (MEA 2006:133)

[142] MEA 2006:131.

[143] MEA 2006:182.

[144] MEA 2006:304, Table 9.2.

[145] MEA 2006, Table S3.

[146] MEA 2005, Table S2.

[147] MEA 2006:155.

The groups attempted to check for consistency between quantitative and qualitative scenario elements through periodic consultations, particularly for feedbacks that could not be modeled analytically. Some of these were interactions between direct drivers and ecosystems, but the most difficult occurred in scenarios that assumed strong socio-economic feedbacks and regulating mechanisms. Adapting Mosaic, for example, assumed strong feedbacks from new ecosystem observations and knowledge to changes in human behavior that could not be incorporated into the models used. Representing these required allowing qualitative scenario logic to override both the quantitative results and the structure of models. Unfortunately, time limits prevented this consistency checking from being done thoroughly, so unexamined disparities between qualitative and quantitative aspects of the scenarios remained a significant weakness.

Many of the conclusions developed from the scenarios are common to all four scenarios, while in others Order from Strength is the exception. For example, one major conclusion is that rapid conversion of ecosystems for use in agriculture, cities, and infrastructure will continue, and that habitat loss will continue to contribute to biodiversity loss. However, if ecosystem services increase as projected, some ecosystem services – although not biodiversity – may be decoupled from ecosystem stresses. Food security is projected to remain out of reach for many people. Extreme, spatially diverse changes are projected for freshwater resources, with general deterioration in developing countries under both "reactive" scenarios. Increasing demands for fishery products are projected to increase risks of regional marine fishery collapses.[148] In sum, ecosystem services show mixes of improving and worsening trends in all scenarios except Order from Strength, in which nearly all ecosystem services are projected to be more impaired in 2050 than in 2000. The same three scenarios also suggest that significant changes in policies, institutions, and practices can mitigate some negative consequences.[149]

In sum, the MEA scenarios project invested substantially more effort in developing rich qualitative and narrative scenarios than the SRES, but also fell short on integrating qualitative and quantitative components. In part because of the greater elaboration of the qualitative components, this limited coordination resulted in significant inconsistencies and requirements to resolve conflicts between the two components. These inconsistencies arose even with just one model used for several components of the assessment, so the challenges of harmonization among models – and the associated possibility to explore model-structure uncertainty – did not arise. A related problem was that for many factors it was difficult to generate the desired level of variation between scenarios.[150] This raises issues of potential methodological interest, such as how to distinguish robust results from inadvertent convergence of assumptions or model structures, which remain to be investigated. Finally, the great breadth of conditions represented in the scenarios, as well as possible concerns with logical circularity between their presumptions and results,[151] make interpreting the significance of the results difficult.

The experience of this scenario exercise provides a different perspective on some of the same key challenges for future scenarios highlighted by the other activities reviewed. The quite distinct difficulties faced here in attempting to combine quantitative and qualitative scenarios highlight the central importance and the difficulty of developing new methods to integrate these two approaches. In addition, this experience highlights the value of clarity about the intended uses of scenarios, including clarity about whether they are intended to address specific questions, guide decisions, or explore long-term conditions. The risk of scenarios becoming

---

[148] MEA 2006, Table S3.

[149] MEA 2006:127.

[150] Report of the First Meeting of the MEA Global Modeling Group, 7 Jan 2003, at www.usf.uni-kassel.de/ma-gmgroup/dl/first_report.doc; Second Report of the MEA Global Modeling Group, 7 March 2003, available at www.usf.uni-kassel.de/ma-gmgroup/dl/sanjose_report.doc.

[151] This concern is particularly present regarding implications of the assumption that ecosystem management is either proactive or reactive (See, e.g., MEA 2006, Ch 8.4.2.1 and Ch 9).

less useful due to breadth and vagueness may be particularly acute for scenarios that attempt to capture multiple stresses on some system – even though such multi-stress assessment is repeatedly advocated for climate-change and other forms of environmental assessment.[152]

---

[152] NAST 2001.

Challenges and Controversies in Scenarios for Climate Change

This section discusses several issues that have arisen in multiple scenario exercises related to climate change, issues that pose challenges for expanding the usefulness of scenarios to climate change analysis, assessment, and decision support. Section 4.1 examines the type of information needs of specific types of decisions related to climate change and considers the requirements and challenges of crafting scenarios to serve these needs. Section 4.2 considers the use of scenarios that has been more common thus far, in structuring climate-change assessments and framing broad policy debates, and identifies the distinct challenges in enhancing the value of scenarios in these purposes. The remaining sub-sections examine particular design challenges in crafting scenario exercises: how to structure interactions between experts and stakeholders in developing scenarios; how to communicate scenarios to potential users not involved in their creation; how to pursue the two, not perfectly aligned goals of consistency and integration in scenarios; and how to represent and interpret uncertainty in scenarios. Throughout this section, we present illustrative examples of scenario activities in text boxes. These examples shed additional light on various challenges, especially relating to scenarios' use in decision-making.

4.1. SCENARIOS AND DECISIONS

As discussed in Section 1, the general purpose of scenarios is to inform decisions, but their connection to specific, identified decisions can be more or less close and direct. In interpreting and evaluating present experience with scenarios and identifying key challenges in making them more useful, it is important to distinguish scenario exercises by their major characteristics, including their specificity, their proximity to decisions, the degree of normative presumptions embedded in them, and where they lie in the causal chain outlined in Section 2. To consider how scenarios can help inform climate-change decisions, we must first specify the relevant decisions and decision-makers more sharply. This section considers the major concrete decisions that comprise a response to climate change. Decisions related to assessment, modeling, and research are considered in Section 4.2. This discussion must be somewhat hypothetical, extending from rather thin current practice to reasonable speculation about future decisions and likely information needs.

Because the dynamics of climate change operate on multiple spatial scales from the local to the global, there is no single global climate-change decision-maker. Rather, many distinct decision-makers with diverse responsibilities will affect and be affected by climate change. Because of climate's recent appearance on policy agendas and its dense connections to other issues, many of these decision-makers' primary responsibilities are defined as something other than climate change. Some of them are already considering how climate change might affect their responsibilities, but many are not.

Section 2 described climate-change decisions using the conventional dichotomy of mitigation versus adaptation. To consider potential contributions of scenarios in more detail, we discuss three types of decision-maker: *national officials*, *impacts and adaptation managers*, and *energy resource and technology managers*. These can often be identified as particular programs, divisions, agencies, organizations, or individuals, each with different responsibilities and types of information they might consider in making their decisions. All three groups face decisions under uncertainty with long-term consequences related to climate change, and so might benefit from scenarios providing structured information and assumptions about the values at stake, the available choices, and their consequences under alternative climate-change futures.

National officials' responsibilities are the broadest and the most likely to be explicitly related to climate change. They develop national policies that target greenhouse-gas emissions and motivate investment in technologies that will influence future emissions trends. They negotiate policies internationally with officials from other nations, and with sub-national officials who may share mitigation responsibilities or undertake mitigation measures at their own initiative. They also have responsibilities to anticipate and respond to climate-change impacts in their nations. Their climate-change responsibilities are open-ended, not limited to mitigation and adaptation: these decision-makes will determine the extent to which other responses such as geoengineering are considered, and the design of systems and institutions for assessment. They are also responsible for overall national welfare, including not just the environmental effects of their decisions but also other linked national interests such as economic prosperity and security.

Impacts and adaptation managers have responsibility for particular assets, resources, or interests that might be sensitive to climate change. They must decide how to anticipate, prepare for, and respond to the threat, minimize its harm, and maximize any associated benefit. They may be private or public actors – e.g., owners or managers of long-lived assets such as ports or water-management facilities; managers of lands, forests, or protected areas; emergency

preparedness or public health officials; officials making zoning or coastal development policy; or firms in insurance or financial markets who may bear secondary risks from impacts or seek to develop new instruments to exchange these risks. Unlike national officials, these actors' decisions are purely *responses* to climate change, realized or anticipated: they have little influence over how the climate will change. Their responsibilities will often connect with the impacts-related responsibilities of national officials, but are narrower in scale or scope. Impacts and adaptation managers would be concerned not with aggregate climate-change impacts on the United States, but with more specific impacts such as those on seasonal flows and water-management operations on the Upper Mississippi.

Energy resource and technology managers include developers and operators of fossil or non-fossil energy resources, investors in long-lived energy-dependent capital stock such as electrical utilities, and researchers, innovators, and investors in new energy-related technologies. These decision-makers are mostly but not exclusively in the private sector. Their decisions may have consequences that interact with various processes operating over multiple time-scales, from short-term market responses, to decadal-scale processes of investment, resource development and depletion, and penetration of new technologies, to century-scale processes of climate change.[153] These actors' decisions will strongly influence society's ability to control greenhouse-gas emissions. This group also includes energy consumers such as firms or public agencies considering mitigation actions in their own operations. While their areas of responsibility may be vulnerable to climate change and its impacts, the largest climate-related risks for this group are likely to come not from climate change itself, but from climate-change policies: national mitigation policies, and other market and regulatory decisions that shape the outcomes of private mitigation activities.

At greatly varying levels of precision and specificity, scenarios can present two types of information to support decisions by these three types

---

[153] Shell International 2001, Davis 2003.

of actors. Scenarios can represent potential future developments that may threaten decision-makers' interests or values, call for decisions, or challenge conventional thinking and practices. And they can provide a structure to assess potential consequences of alternative decisions for things that matter to the decision-maker. Beyond this generalization, the three types of decision-makers will differ substantially in the specific types of information they need, the time horizons of their decisions, and the type and extent of causal connections between their decisions and the conditions specified in scenarios.

4.1.1. Scenario needs: national officials

As national officials have the broadest responsibilities related to climate change, they are also likely to have the broadest information needs. In their responsibilities to build national adaptation capacity and manage key vulnerabilities, their needs are similar to those of impacts and adaptation managers: scenarios of potential future climate change under specified emissions assumptions and resultant impacts on particular resources and communities in their nation, with particular focus on areas of greatest vulnerability. They will likely have less need for fine spatial and sectoral detail in potential impacts, but more need for consistent scenarios that allow comparison and aggregation across sub-national regions and sectors. These will help to prioritize, identify key areas of vulnerability, and estimate aggregate costs for planning purposes.

In their responsibilities for national mitigation policy, national decision-makers will also need information about the aggregate impacts of climate change, since the more severe climate impacts are likely to be, the greater the justification and likely political support for mitigation measures. But mitigation decisions also require additional information – including projections of future emissions in the absence of explicit mitigation efforts, and the consequences of alternative mitigation policies, in their effects on emissions, their cost, and their implications for other dimensions of national interest.

These needs introduce a dimension of complexity into mitigation scenarios, sometimes called "reflexivity." Because mitigation policies seek to reduce future emissions by altering the socio-economic drivers of emissions growth, the analysis of mitigation policies and their consequences must be coupled to the causal logic of emissions scenarios. Whereas climate scenarios can be treated as exogenous when assessing adaptation decisions, emissions scenarios cannot be treated as exogenous in assessing mitigation decisions. Any emissions scenario embeds some assumptions about mitigation policies, assumptions that may have to be changed to assess particular mitigation policies. This effect will be strongest when emissions projections and mitigation options are being considered at the same spatial scale, e.g., national mitigation policies are being assessed relative to national emissions projections. The effect of national mitigation strategies on global emissions will be weaker: no nation controls global emissions trends, and the effects of small nations' mitigation strategies on global trends can be very small.

Scenarios to inform mitigation decisions are also likely to require considering alternative assumptions about the policy context in which these decisions are made. The effects of national mitigation strategies – including how much they reduce national emissions, as well as their costs and other consequences – will depend on the economic, technological, and policy context, including related decisions by other major nations, individually and through international coordination. Assumptions about the policy context will be less important in scenarios to inform international mitigation decisions, since when decisions are globally coordinated there is no "elsewhere" – but alternative assumptions about nations' degrees of compliance and form of implementation of international commitments may still be needed.

Scenarios of emissions, climate change, and impacts inform mitigation decisions by helping to characterize the potential severity of climate change and therefore how important it is to control emissions. This support is indirect, serving primarily to elevate or moderate the general level of concern on the issue. More focused work on mitigation has been done using con-

> Any emissions scenario embeds some assumptions about mitigation policies, assumptions that may have to be changed to assess particular mitigation policies.

structed scenarios of limited emissions, often aiming at stabilizing atmospheric concentrations or radiative forcing at various levels, and examining the configurations of technology, energy resources, and economic and population growth that are consistent with the specified scenario. Some studies have used quantitative models to estimate costs of such scenarios, relative to an assumed baseline emissions scenario.[154]

4.1.2. Scenario needs: impacts and adaptation managers

To assess the threats and opportunities they face and evaluate responses, impacts and adaptation managers need scenarios of potential future climate change, its impacts in their areas of responsibility, and the factors that influence vulnerabilities. With few exceptions, these actors' decisions will have no effect on the climate change to which they must respond, so scenarios of climate-change stresses can be constructed independently of the assessment of potential adaptation decisions, without concern for feedbacks that may modify the conditions specified in the scenario.

Particular decision-makers' needs will be highly specific in the variables they require, and their time and spatial scale and resolution. A planner of water-management infrastructure may need monthly or finer-scale rain and snow projections over a watershed; a designer of coastal infrastructure may need probabilistic projections of sea level, storm intensity and frequency, storm surge, or saltwater intrusion. But in their climatic elements, these information needs all rest on a common core of scenarios of global climate change and emissions drivers. This dual structure of information – highly particular climate variables, based on a set of common "core scenarios" – suggests a cross-scale organizational structure for providing scenario information: commonly produced scenarios of climate change and other components requiring consistency, specialized expertise, or high-cost resources; development of decentralized capabilities in impact assessments to adapt these core scenario elements and develop assessment-specific extensions; and close communication between these groups to ensure that useful vari-

ables are generated and saved, and that data and documentation are transferred accurately.

This is the area of climate-related decisions for which the provision of information from climate-change scenarios is most advanced. Still, further progress is needed in the development and use of scenarios of socio-economic conditions, and in creation of methods and tools to augment centrally provided scenario information with information tailored to specific impact assessments. In addition, many impacts-related decisions will require scenarios of climate change in the context of other linked stresses and changes.

4.1.3. Scenario needs: energy resource and technology managers

Energy and technology managers will most benefit from scenarios that explore alternative policy regimes and their consequences for the value of energy and technology assets and investments. For some, the predominant concern may be overall policy stringency, perhaps summarized as alternative emissions-price trajectories over time; for others, specific details of policy design and implementation may need to be considered. Scenarios of emissions, climate change, and impacts only matter for decisions via their likely influence on policy stringency, and so do not need to be explicitly represented in scenarios. These actors may have some influence on policy, but probably not such strong influence that climate-policy scenarios would have to incorporate feedbacks from their own advocacy efforts.

Unlike the other two types of decision-makers, these actors are likely to compete with each other. If, for example, they are investors allocating research effort between higher and lower-emitting energy sources, those who better anticipate future policy will benefit relative to those who do worse. If they use scenarios, they may consequently choose to produce them privately, perhaps coupled with other analyses to generate practical guidelines for investments.[155] As for the other types of decision-makers, these specialized scenarios could be based on general scenarios of global emissions and climate

---
[154] Morita et al. 2001, CCSP 2007.

[155] Ged Davis, personal communication. (posted expert review comments).

change. Published scenarios produced to date on the climate-change issue, however, have not considered mitigation policies with the specificity necessary to inform these actors' decisions.

4.1.4. Representing decisions in scenarios

A major challenge to developing scenarios to support decisions is reflexivity, that is, how to represent decisions within scenarios without making scenarios either circular or contradictory. In meeting this challenge, the most basic distinction to draw is between decisions by the scenario's targeted users and decisions by other actors. From the users' perspective, decisions by others over which they have no influence are indistinguishable from non-choice events. If the factors influencing these decisions are confidently understood, they might be represented deterministically, like well-understood biophysical or economic processes. In the more frequent case that others' choices cannot be confidently predicted, they might be represented as uncertainties – again, just like uncertain biophysical or economic processes. As with all uncertainties, how to treat them depends on their judged importance for the users' decisions: if it is high, they can be represented in alternative scenarios; if not, they can be fixed at some best-guess value for all scenarios. In either case, these decisions are treated exogenously.

Representing decisions by the scenario users is fundamentally different. Since the scenarios are intended to inform these decisions, alternative choices should not be represented as exogenous uncertainties but be stipulated independently from the scenarios. Users can then explore their implications under conditions imposed by scenarios, including representation of major uncertainties. Various degrees of coupling can be required between the logic of scenarios and the analysis of consequences of the users' decisions. In scenarios for impacts, these can usually be separate; in scenarios for mitigation, they may have to be closely coupled, since emissions scenarios may change under alternative assumptions about mitigation decisions.

In scenarios to inform global climate-change decisions, the sharpest question is how to represent mitigation decisions within scenarios.

Following the general reasoning above, how these are treated should depend on what type of decision is being informed. In climate scenarios to inform impact assessments and related decisions, the scenario users are not considering mitigation decisions and have little influence over them, so emissions scenarios should include assumptions about the likely or plausible range of mitigation efforts. The range of future climate change considered may thus be narrowed to reflect the possibility of negative social and political feedbacks: sustained rapid emissions growth may generate pressure for aggressive mitigation, due to increasing signs of climate change, alarming projections of future change, or other environmental harms from rapid expansion of coal or synthetic fuels.

Such a negative-feedback mechanism may not be effective, of course. Many factors could intervene: mitigation measures may not gain enough support to be adopted, socio-political capacity to enact stringent policies may be diminished, policies adopted may be ineffective, or early technology or policy decisions may lock in high-emitting future paths. But to the extent that such a negative-feedback mechanism does operate, persistence of the highest emissions paths beyond a few decades would become unlikely.

Parallel reasoning may apply to extremely low emission paths, if sustaining such low emissions requires continued costly mitigation efforts that come to be seen as unnecessary. This negative-feedback mechanism would likely be weaker than that operating at the high end of the emissions distribution, however, because long time-constants mean that increasing signs of climate change are likely to continue through most of the 21st century even if we follow a low-emissions path. If impacts assessors and managers judge these negative feedbacks to make extreme emissions paths sufficiently unlikely, particularly high ones, they may reasonably decide not to consider these extreme emissions futures in their planning for adaptation.

For scenarios to inform mitigation decisions, particularly at the international level, the situation is different. Informing these choices requires information about potential emissions paths and their consequences under all levels of

In climate scenarios to inform impact assessments and related decisions ... emissions scenarios should include assumptions about the likely or plausible range of mitigation efforts.

mitigation effort that decision-makers might plausibly consider – including no additional measures, or even reversal of previous measures if this is on the agenda. Consequently, in contrast to scenarios for impacts, extreme emissions futures should not be excluded when assessing mitigation decisions. For example, if scenarios that truncate high-emissions futures by assuming stringent mitigation are used to support a decision that stringent mitigation is not necessary, the result is contradictory: a conclusion supporting a decision is based on the presumption of the contrary decision. To avoid such contradictions, scenarios to inform mitigation decisions must consider alternative mitigation choices explicitly, not embed them implicitly in the underlying logic of the scenario.

Moreover, national officials act only for their own nations in the near term, even when they negotiate global mitigation. They may make choices for long-term planning and institutional design for future mitigation as well, but it is their successors who will decide whether to continue, strengthen, or otherwise change measures adopted today. From the perspective of current national officials, mitigation decisions by other nations and in the future fall between the two cases discussed above: they are not controlled by the scenario user, but can be influenced to some degree. For policy choices by other nations, national officials may need to be advised in two modes, reflecting their dual responsibilities to make national policy and to negotiate international agreements. In the latter capacity, alternative approaches to global mitigation strategy should be represented as choices. But when they consider national decisions separate from globally coordinated strategy, relevant decisions of other major nations should be represented as uncertainties. This may require use of two distinct types of scenarios to advise development of different aspects of national mitigation policy.

How to represent future mitigation decisions poses a still harder dilemma. On the one hand, it appears risky or even irresponsible to assume that the bulk of mitigation efforts can be left to future decision-makers, even if we assume this will be easier for them because of greater wealth or technological prowess. On the other hand, assuming that future decision-makers cannot be

relied on to act responsibly at all can easily lead to decisions that incur excessive costs, by trying to achieve rapid mitigation immediately or tie future decision-makers' hands.

Two approaches appear promising for integrating future mitigation decisions into scenarios to inform current decisions. Scenarios could presume that today's decision-makers choose the future path of mitigation, allowing them to assess and contribute to a trajectory of effort that considers the welfare of both current and future citizens. Alternatively, scenarios could treat future large-scale mitigation choices as uncertainties represented in alternative scenarios, while also considering how current choices can seek to influence the opportunities and incentives faced by future decision-makers.

In sum, the importance of connecting scenarios to actual decisions is widely recognized, but there is a large gap between, on the one hand, the value scenarios could provide to climate-change decisions and the aspirations of scenario producers to provide that value, and current practice on the other hand. There has been little use of scenarios to directly inform climate-change related decisions, although there appears to be a sharp increase in the interest of decision-makers and early attempts. The rapid increase in interest is particularly evident for informing decisions related to climate-change impacts and adaptation. There are fewer indications of similarly direct use of scenarios to inform mitigation decisions, perhaps in part because nearly all current mitigation decisions have been near-term.

Mitigation decisions at the national and international level have taken scenarios into account indirectly. Most scenarios have been constructed to provide inputs to assessments, models, or other analyses. This has included serving as inputs to the production of other types of scenarios, which then describe other potential future conditions that depend on those specified in the scenario, as for example a model-based climate scenario depends on inputs from an emissions scenario. While these uses can be characterized as supporting decisions (i.e., decisions about assessments, modeling, and research), their connection to concrete decisions of mitigation and adaptation is indirect, achieved through contributions such as sup-

...in contrast to scenarios for impacts, extreme emissions futures should not be excluded when assessing mitigation decisions. ... Scenarios to inform mitigation decisions must consider alternative mitigation choices explicitly, not embed them implicitly in the underlying logic of the scenario.

porting strategic planning and risk assessment, providing advance analysis for potential future decisions, exploring plausible extreme cases, helping to characterize and prioritize key uncertainties, or educating decision-makers or the public. This description applies to the major scenario exercises discussed in this report, including the IPCC emissions and climate scenarios, the US and UK assessments of climate impacts, and the MEA scenarios.

BOX 4.1. Scenarios for Climate-Change Adaptation in the New York Metropolitan Region

Three linked activities – the Metropolitan East Coast (MEC) assessment of the US National Assessment, the New York Climate and Health project (NYCHP), and the New York City Department of Environmental Protection (NYCDEP) Task Force on Climate Change – have used or are using scenarios to assess impacts of climate change on the New York Metropolitan Region, identify areas of vulnerability, and inform regional planning and decision-making.[156]

The MEC assessment began with a regional workshop in April 1998, involving about 150 participants, including public agencies at all levels of government as well as climate researchers. The subsequent assessment was conducted by sector teams of researchers and officials from agencies responsible for the study sectors. Teams developed regional scenarios of climate change and sea level rise based on the downscaled climate-model scenarios provided by the US National Assessment, plus two additional scenarios based on extrapolation of recent regional climate trends and historical extremes. The scenarios were used to project climate-change impacts on beach nourishment, 100 and 500-year flood heights, wetland aggregation and loss, adequacy of the water supply system under droughts and floods, illnesses from acute air-pollution episodes, and peak energy loads. These impact projections were used for preliminary assessment of adaptation strategies and policies.

Following the MEC Assessment, the NYCHP created updated regional climate scenarios in consultation with an expert-stakeholder Advisory Board. This study further analyzed public health impacts, focusing on air quality and extreme heat events. The updated climate scenarios used the IPCC A2 and B2 emissions scenarios driving global and regional climate models to create downscaled scenarios for the region. These were augmented with newly developed scenarios of future regional land use and population growth based on the IPCC A2 and B2 storylines.

In response to the widespread public attention received by the MEC Assessment Report, the Commissioner of the NYCDEP established the Climate Change Task Force, a collaboration among regional researchers and the agency that manages the water system. The Task Force uses the latest climate-model simulations from the IPCC Fourth Assessment Report, as well as additional global and regional climate models, to develop new regional scenarios. These will include probability distributions of average and extreme temperature and precipitation change, as well as sea level rise. The Task Force is also developing qualitative scenarios of extreme sea level rise in the region. DEP is using these results to develop a comprehensive adaptation strategy for the New York City water system, including assessment of many specific adaptation options, that considers both uncertainties in future climate change and managerial factors such as the time horizon of different adaptation responses and capital turnover cycles.

This is a successful example of scenario-based assessment of climate impacts and adaptation options. The scenarios are connected with the concrete responsibilities and concerns of stakeholders, who were involved in their design from the outset. Although officials have found the wide range of uncertainty in climate scenarios difficult to incorporate into infrastructure design specifications, particularly for precipitation, the exercise has effectively conveyed the challenges posed by future climate uncertainty to current decisions of planning and infrastructure design. Stakeholders' willingness to support and participate in three separate phases of these activities and NYCDEP's incorporation of them into a strategic planning exercise provides clear evidence of the practical utility of the exercises.

---

[156] Rosenzweig and Solecki 2001; Kinney et al. 2005, 2006; Rosenzweig and Major 2006.

BOX 4.2. Scenarios of Sea Level Rise along the Gulf Coast

Sea-level rise is one of several factors that contributed to the decline of coastal ecosystems along the US Gulf of Mexico coast in the 20th century (Figure 4.1).[157] In southeastern Louisiana, where the local rate of land surface subsidence is as high as 2.5 cm per year, rise in local "relative sea level" may be the most important factor in the rapid loss of coastal zone wetlands over the past several decades.[158]

Despite the importance of sea level rise in historical losses of coastal lands, planning projections of future changes in coastal Louisiana used by both federal and state agencies prior to the devastating impact of Hurricanes Katrina and Rita in 2005 were based on just one scenario: no change in the rate of sea level rise. No alternative sea level scenario was considered in the plans then being developed to restore and protect the Louisiana coastal zone.[159] This assumption sharply contrasts with the IPCC projections, which state that the global average rate of sea level rise in the 21st century may increase 2- to 4-fold over that of the 20th. Such increases will exacerbate wetland losses throughout the Gulf Coast region and obstruct restoration plans that do not take account of likely increases in water levels and salinity.

The ecosystem modeling team working for the State of Louisiana and the US Army Corps of Engineers is presently integrating accelerated sea level rise scenarios into planning exercises that will aid federal and state agencies in evaluating restoration alternatives.[160] The State of Louisiana is consulting with the Rand Corporation to obtain probability estimates for various scenarios of sea level change to help guide engineering decisions and the design of projects aimed at restoring levees and coastal landforms that protect coastal communities.[161] Sea level rise scenarios are also being used to assess the impacts of climate change and variability on the Gulf Coast transportation sector. To assess transportation impacts, a sea level rise simulation model developed by the US Geological Survey generates scenarios of sea level change using over a dozen GCMs and six SRES emission scenarios.

Sea level rise scenarios are important not just in regions like Louisiana. The Big Bend region of the Florida panhandle is experiencing very little vertical movement of the land surface, so sea level there has been rising at approximately the global average rate of 1 to 2 mm per year. But even here, coastal wetlands positioned on flat limestone surfaces may be subject to highly nonlinear effects as sea level reaches a threshold at which large areas are subject to increased salinity or inundation.[162]

Regional scenarios of potential sea level rise are needed to support coastal management and protection activities, as well as plans for wetland restoration and post-hurricane reconstruction. Absent consideration of such scenarios, restoration and rebuilding programs are likely to lock in errors that result in wasted resources and avoidable increases in future vulnerability.

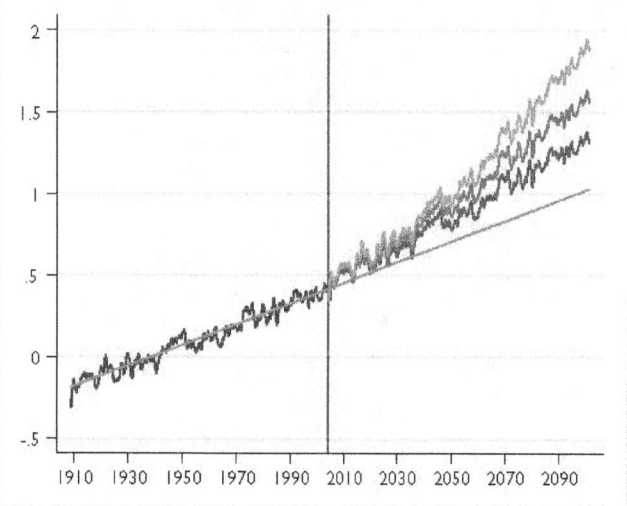

Figure 4.1. Output from a Gulf Coast sea level rise scenario tool

Historical sea level change and projected sea level rise under three greenhouse-gas emissions scenarios, in meters, are shown for Galveston, Texas. Both historical data and future projections are smoothed from monthly data using a 12-month moving average. *(Source: Thomas W. Doyle, National Wetlands Research Center, United States Geological Survey.)*

— Historical Data
— Low (B1) Scenario
— Middle (A1B) Scenario
— High (A1FI) Scenario
— Linear Trend from 1909–2003 Monthly Data

[157] Gosselink 1984, Williams et al. 1999, Burkett et al. 2005, Morton et al. 2002.

[158] Shinkle and Dokka 2004, Barras et al. 2003.

[159] US Army Corps of Engineers 2004.

[160] See, e.g., http://www.clear.lsu.edu/clear/web-content/index.html.

[161] Presentation by Randy Hanchey, Louisiana Department of Natural Resources, to Governor's Advisory Commission on Coastal Protection, Restoration and Conservation, Baton Rouge, LA, June 22, 2006.

[162] Cook 1939, Doyle et al. 2003.

BOX 4.1.3. Scenarios in the California Water Plan

The California Department of Water Resources (DWR) produces an updated California Water Plan every five years. The plan projects water supplies and demands, and evaluates current and proposed demand-management programs and supply investments, to "provide a framework for water managers, legislators, and the public to consider options and make decisions regarding California's water future."[163]

In contrast to prior plans that constructed only one future scenario, the 2005 plan explicitly considered uncertainty in supply and demand projections. Three alternative scenarios of supply and demand conditions were constructed through 2030: one extending current trends in population and economic growth, agricultural production, environmental restrictions on water use, and water conservation occurring without policy initiatives (e.g., through equipment replacement, technological change, and revised building codes); and two presenting higher and lower increases in demands. The report of the 2005 plan discusses global climate change and the potential challenges it poses to water supply and demand in California, but climate change is not explicitly represented in the plan's three scenarios.

In addition to adopting these scenarios, the State of California is developing data and analytic capacity to enrich the treatment of uncertainty and climate change in future plans. In parallel with development of the three principal scenarios in this plan update, DWR sponsored development of several analytic tools to strengthen the treatment of uncertainty in future plans. In addition, the California Climate Change Research Center with co-sponsorship from DWR is developing fine-scale regional climate-model scenarios to support analysis of climate-change impacts on water resources.[164] The DWR plans to incorporate these climate-change scenarios explicitly in the next plan update in 2010.

4.2. SCENARIOS IN ASSESSMENTS AND POLICY DEBATES

Within large-scale assessments of climate change or other environmental issues, scenarios can serve several roles. Most straightforwardly, they can provide required inputs to other parts of the analysis, as the IPCC emissions scenarios support the controlled comparison of climate-model runs. They can also serve as devices to organize and coordinate the multiple components of a large-scale assessment, particularly when much of the assessment is forward-looking. In the IPCC assessments, for example, emissions scenarios have not just been used to drive coordinated climate-model projections, but have also increasingly been followed through to coordinate characterization of climate impacts and adaptation opportunities, and used in a more preliminary way to organize assessments of the economic and technological implications of alternative mitigation strategies.

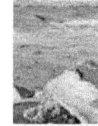

Similarly, the US National Assessment and UK Climate Impacts Programme have both attempted to identify a small set of climate and socio-economic scenarios, to coordinate and gain comparability across multiple studies and allow aggregate assessment of impacts and vulnerabilities at the national level.

In a broad assessment including many teams considering separate questions of climate-change, impacts, mitigation, and adaptation, simple coordinating devices are needed to make teams' work comparable and allow synthesis to produce aggregate conclusions. Emissions scenarios are natural devices to provide such coordination, both because emissions hold the clearest near-term opportunities for intervention, and because they have clear and recognized connections both directions in the causal chain, to every aspect of the climate-change issue. However, in part due to management issues, these efforts to use scenarios as broad coordinating devices have not been wholly

---

[163] California Water Plan home page, http://www.water-plan.water.ca.gov.

[164] California DWR 2005:4-32.

satisfactory in practice. To serve as coordinating devices, scenarios must be developed and disseminated early in the process, preferably before the work of assessment teams even begins. Moreover, they must be documented with detailed information about the process and reasoning used to generate them, including explicit identification of underlying assumptions and supporting data, models, and arguments. In practice, timely, detailed, and transparent dissemination of scenario information has rarely been achieved.

Scenarios used in large-scale assessments can also make other contributions that are related to the prominent dissemination a major assessment receives. They may, for example, be used as inputs to planning or decision-support processes that were not part of the original assessment. In such use, they may gain a more direct connection to decision-making than they had in their original production or use. Scenarios of global emissions and the model-based climate scenarios based on them especially lend themselves to such derivative uses, informing many different decisions by diverse actors.

Scenarios in prominent assessments can also contribute to the framing of public and policy debates. In this role, scenarios inevitably become political objects, in two senses. They are subject to political forces that seek to influence their development, and political reactions to them once developed. These pressures pose challenges and risks that differ quite markedly from those that apply in using scenarios to inform decision-making, where we tend to assume a greater degree of commonality of knowledge, perspective, and interest in the process among participants and some group of relatively well-defined users.

Within scenario exercises, various actors may seek to bias scenarios' content to help advance their policy preferences or their broader political objectives, by limiting consideration to futures they judge desirable or showing some problem in an acute state that would appear to demand a response. While it is not possible to eliminate biases in scenarios, unacknowledged

normative biases in scenarios can pose the risk of excluding consideration of futures that are judged undesirable or that pose sharp decision-making challenges. Such biases can be difficult both to detect and to correct. Beyond exhorting developers to scrutinize scenarios critically to avoid bias, the best protection against such biases lies in transparency about the assumptions and information underlying scenarios and associated judgments of likelihood.

Other political pressures come onto scenarios in the broader use, debate, and criticism that they encounter after release. For impartial support of policy decisions, scenarios should fairly present knowledge and uncertainty about potential variation on important dimensions. This typically requires consideration of a wide range of potential futures – often a wider range than relevant decision-makers might initially think plausible, due to well-known habits of conventional thinking and excessive confidence.

Sometimes a scenario's implications for decisions may be obvious. For example, a scenario might represent developments so severe that most people would judge it to demand intervention. Another might represent developments that most people would judge inconsequential or beneficial, so not meriting any intervention. A wide-ranging set of scenarios may include examples of both such extremes. Consequently, such a wide range of potential futures in a set of scenarios – even if this is faithful representation of present knowledge and uncertainty – provides opportunity for partisan distortion and efforts to make scenarios policy-prescriptive.

In global change scenarios, conflicts and opportunities for bias arise most acutely over emissions scenarios. Since much of the uncertainty about climate change beyond 2050 comes from uncertainty in future emissions trends, actors with strong policy preferences can highlight emissions scenarios that lend support to their views. Those who advocate aggressive mitigation may highlight the highest-emissions scenarios to emphasize the elevated risk of climate change that would follow. Those who oppose mitigation may highlight the lowest-

Scenarios inevitably become political objects, in two senses. They are subject to political forces that seek to influence their development, and political reactions to them once developed.

emission scenarios to suggest that no action to limit emissions is warranted. Because scenarios are used when knowledge of causal processes is weak, it is easy to make any scenario appear salient and likely, even if it is extreme. It is equally easy to probe inside the details of any scenario to find inconsistent or implausible implications, particularly when a scenario is rich in detail.

But, although political actors may have legitimate reasons to highlight one extreme scenario, it is not appropriate for any scenario to dominate assessment or consideration of decisions. A claim that only a single scenario is plausible – especially one near the top or bottom of the present range – is a claim to predict the future, which can be readily dismissed. Claims that a particular scenario is *implausible* cannot be so easily dismissed, however, since scenarios represent only the imperfect judgment of the team that produced them. Leaving aside scenarios that violate clear principles of science (e.g., one whose energy assumptions violate the laws of thermodynamics) or economics (e.g., one that presumes a large new capital stock in a few decades without the investments needed to create it), it is possible to construct pictures of the next century so extreme or unprecedented that most observers would agree they do not merit serious consideration. But short of such an extreme – which describes no global-change scenario discussed here or known to us – claims that a broad class of potential futures is implausible should have to pass a high hurdle. Identifying specific extreme or implausible elements within a scenario does not suffice to make this case, since virtually any scenario will be found to contain these if scrutinized closely enough. Nor does identifying ways that a scenario of future change diverges from some established trend or pattern, since established trends can and do change.

Historical studies of forecasting exercises such as energy forecasts have repeatedly found them too confident that the future will extend recent trends.[165] The threshold any single scenario must pass is to appear sufficiently plausible or instructive to merit consideration in planning and analysis, and this is a judgment to be made by developers and users – with enough transparency about underlying assumptions and reasoning that users can make an informed judgment. A set of scenarios should be constructed so that the range of conditions they represent encompasses present knowledge and relevant uncertainties that might influence mitigation or adaptation decisions. Since subjective judgments cannot be avoided in constructing scenarios, the range provided should err on the side of being broad rather than narrow, at least initially. Identifying problems with one scenario does not necessarily impugn the credibility even of that one scenario, certainly not the whole set, because scenarios cannot be consistent in every underlying detail.

In subsequent revisions as knowledge advances, scenarios can continue to play their role coordinating assessments and framing policy debates with more focus and less arbitrariness. Continuing research and analysis might come to identify some scenarios as severe in their consequences and others as inconsequential, or might revise the initial characterization of the determinants and feasibility of particular scenarios, including suggesting that some are too unlikely to merit serious consideration. These judgments can feed into decisions about continuing analysis of scenarios: which ones can be dropped and what new ones should be added. One major basis for updates in scenarios will be policies adopted, which can set a baseline to focus further deliberations. Perfect attainment of targets and success of policies should not be assumed, but scenarios can focus subsequent debate by posing such questions as "What if we just meet this target; what if we fall short by this much; and what if we exceed it by this much, or adopt these additional measures?"

A claim that only a single scenario is plausible – especially one near the top or bottom of the present range – is a claim to predict the future, which can be readily dismissed.

---

[165] Smil 2005, Greenberger et al. 1983.

BOX 4.4. Scenarios of Ozone Depletion in International Policy-making[166]

Emission scenarios of chlorofluorocarbons (CFCs) and other ozone-depleting chemicals substantially influenced policy debates over control of these chemicals to protect the ozone layer. Until the early 1980s, these policy debates used a convention to project ozone losses that originally served as a simplifying research assumption: constant emissions forever. This convention has obvious research benefits, like the simple doubled-CO_2 equilibrium scenarios used in climate models. It standardized input assumptions, allowing exploration of scientific and modeling uncertainties without the confounding effect of different emissions assumptions. Moreover, this convention made no claim to realism, and so avoided distracting arguments over whether one emissions projection or another was more realistic. Nevertheless, the resultant calculations were frequently mistaken for projections of realistic future trends.

The question of what future emissions trends were likely only became prominent in policy debates around 1983. World CFC production had dropped nearly a third in the late 1970s due to both regulatory and market-driven reductions in their largest use, aerosol spray propellants, and declined further in the early 1980s recession. It was widely argued that further restrictions were unnecessary—CFCs' major markets were saturated and further growth was highly unlikely. The resumption of sharp growth in 1983 undermined this claim, making it clear for the first time that managing the ozone risk required considering scenarios of CFC growth as well as steady-state and decline. How much emissions might grow and what that would mean for the atmosphere remained highly controversial, however.

Emissions of other chemicals complicated the picture. Advances in stratospheric chemistry showed that future ozone loss depended not just on CFCs, but also on emissions of several other gases, including carbon dioxide, methane, and nitrous oxide. But the knowledge and computing capacity to credibly model interactions among all these pollutants only began to appear in the early 1980s. In 1984, a major scientific assessment conducted the first standardized comparison of multiple stratospheric models using a few simple scenarios of emissions trends for CFCs and other chemicals. This exercise had the striking result that under a wide range of trends in other emissions, constant CFC emissions would lead to only very small ozone losses, while CFC growth above about 1 percent per year would lead to large losses.

This result, together with resumed growth in CFC production, had a powerful effect in breaking the deadlock in international negotiations that had persisted since the mid-1970s. Although not the only factor that mattered, this result was crucial in persuading long-standing opponents of CFC controls to accept limits on their future growth. This decisively shifted the agenda for the subsequent negotiations that in 1987 yielded agreement on the Montreal Protocol, which cut CFCs by 50 percent.

In this debate, scenarios used in model-based projections of ozone loss identified divergent trends in future risk that were robust to a wide range of assumptions about trends in other emissions over which there was disagreement. By parsing projected futures into high-risk and low-risk cases, scenarios served to coordinate and simplify a policy debate and so help to focus an agenda for collective decision-making.

---

[166] This example is drawn from Parson 2003.

BOX 4.2.2. Climate-Change Scenarios for the Insurance Industry

The insurance and reinsurance industries face large financial risks from climate change. These are present in many business lines, but the clearest risk is in insurance for property damage from weather-related events, especially windstorms and floods.

In the past two decades, weather-related insurance losses have increased rapidly. By some estimates losses have doubled, even controlling for population and insured value – a much faster increase than that in losses from non-weather events. Climate change is likely to increase insurance risks in multiple ways, increasing the frequency and severity of loss events and also their correlation. Historically based pricing, which is often required by regulations or market conditions, can compound insurers' vulnerability by preventing them from anticipating and adapting to a changed risk environment.

Insurance companies do not use scenarios of future climate change in pricing decisions, because property and casualty contracts are written for short periods, usually one year. Since 1992's Hurricane Andrew, these have mostly been priced using historically based Catastrophic Event Risk Models (Cat models). These estimate losses using a simulated distribution of storm conditions based on historical experience, together with estimates of the durability of insured property. While future climate change poses no risk for these short-term pricing decisions, insurers are concerned that climate change may already have invalidated the historical distributions on which these models are based, by increasing either the probability of severe events or the correlation among them.

Two published exercises have used climate-change scenarios to explore longer-term risks to the insurance industry. The first, conducted for the Association of British Insurers in June 2005, examined potential impacts of climate change on the costs of extreme weather events (both insured and total economic costs) under the six SRES marker scenarios, as well as IS92a and CO_2 stabilization at 550 ppm. The analysis calculated changes in losses due to US hurricanes, Japanese typhoons, and European windstorms.

The second scenario exercise, conducted by Harvard Medical School researchers with sponsorship by Swiss Re and the United Nations Development Program, used two scenarios of 21st-century climate change to examine potential impacts on human and ecosystem health, and associated economic costs, not limited to the insurance industry. The two climate scenarios both assumed CO_2 doubling by approximately mid-century, one with continued incremental climate changes and one with hypothesized nonlinear impacts and abrupt events. The exercise examined potential changes in infectious and water-borne diseases, asthma, agricultural productivity, marine ecosystems, freshwater availability, and natural disasters (including heat waves and floods). The analysis was based primarily on qualitative judgments.

The first scenario showed increased property losses and business interruptions relative to recent trends, emergence of new types of health-related losses, and increasing difficulty in underwriting. The second scenario was qualitatively similar but more severe, with substantial increases in both average losses and variability leading to large premium increases and withdrawal of insurers from many markets, particularly along coastlines. As many insurance firms succumbed to mounting losses, those remaining established strict limits on coverage, shifting more exposure back to individuals and businesses.

Neither of these exercises was connected to any specific, near-term business decision faced by insurance companies. Both could serve longer-term decision-making, however, including planning for reserve accumulation, providing supporting analysis for advocating mitigation and adaptation policies, and supporting changed regulations to allow more flexible pricing of risks experiencing long-term increases. Such exercises can also serve to inform firms' long-term risk-avoidance strategies, including decisions to exit certain areas of business.

BOX 4.6. Scenarios of Climate Impacts in the Columbia River Basin

In conjunction with the US National Assessment, researchers at the University of Washington studied climate impacts on the Columbia River system, which is the primary source of energy and irrigation water for the Northwest and one of the most intensively managed river systems in the world.[167] The project examined the response of annual and seasonal flows both to existing patterns of climate variability – the El Niño Southern Oscillation (ENSO) and the Pacific Decodal Oscillation (PDO) – and to projected 21st century climate change.

The study projected climate change through 2050 using eight climate models driven by one emissions scenario (1 percent per year CO_2 concentration increase), which on average projected 2.3°C regional warming by the 2040s with precipitation increases of 10 percent in winter and a few percent in summer. In the Columbia, these changes are projected to increase flows in winter (because there is more winter precipitation and more of it falls as rain) and to decrease flows in summer (because there is less snowpack and it melts earlier in the spring). The impact of summer decreases is likely to be substantially more serious than that of winter increases. Because the Columbia is a snowmelt-dominated system, winter flows could double or even triple and remain below the present spring peak.

Assessing the impacts of these flow changes requires assumptions about water demands and system management. The study used a reservoir operations model that calculated the combined effects of flow changes and alternative system-operation rules on the reliability of different water-management objectives, such as electrical generation, flood control, irrigation supply, and preserving flows for salmon. Under historical climate variability, all objectives can achieve high reliability in high-flow years (i.e., in the cool phase of ENSO or PDO), but conflict between them occurs in low-flow (warm) years, when only one top-priority objective can be maintained at or near 100 percent reliability and other uses suffer substantial risks of shortfall. Different operating rules distribute this risk among uses.

Under the projected climate and flows of the 2040s, this model showed a pattern of competition between uses similar but additional to what already prevails in low-flow years, suggesting increases in already sharp conflict among uses over flow allocations. One objective could be maintained near full reliability, but other uses suffered reliability losses up to 10 percent from the changed climate, in addition to effects of continued climate variability.

In this analysis, scenarios helped to illustrate interactions between management decisions and climate change and variability, and to explore opportunities and limits for adaptation through management changes alone, with no change in infrastructure or larger-scale policies. This analysis has not been incorporated into any operational decisions, but has been integrated into the Fifth Conservation Plan issued by the Northwest Power and Conservation Council.[168] More detailed assessment of climate-change impacts would require extending this analysis to include projected changes in water demands, both through direct climate effects and through scenarios of regional economic and population growth, allowing a more realistic assessment of potential effects of new water-management investments and changes in large-scale policies to alter water demand, balance competing uses, or improve coordination among the multiple organizations involved in managing the river system.

---

[167] Mote et al. 2003, Payne et al. 2004.   [168] www.nwcouncil.org/energy/powerplan/plan.

4.3. THE PROCESS OF DEVELOPING SCENARIOS: EXPERT-STAKEHOLDER INTERACTIONS

Scenario exercises are collaborative activities that need to be managed. As Section 1 discussed, managing a scenario exercise includes deciding who participates, what jobs they are assigned and how these jobs fit together, how disagreements are resolved, and how much time and money are dedicated to the exercise. These matters can be decisive for the success of an exercise. For some of them, the nature of challenges and tradeoffs they pose are fairly obvious. For example, scenario exercises need enough time to build a team, research scenario components, consult with users, and disseminate results, but often too little time is available, so various compromises must be made. Adding participants expands the expertise and the range of views represented, but increases the time needed for team building and internal communication. Delegating parts of the exercise to smaller groups can overcome this tradeoff, but can introduce coordination problems and inconsistencies among groups. Accepting external direction on scenario exercises increases the chance that decision-makers will take the scenarios seriously, but also increases the risk that they are seen as biased or simply reflect conventional wisdom. These issues pose various challenges, but the challenges are not unique to scenario exercises.

The more central process problems for scenarios concern the relationship between experts and stakeholders in the design, creation, evaluation, and application of scenarios. There has been substantial experience and research in processes for involving stakeholders in environmental decisions, in the United States and other regions.[169] In the most well-established areas of scenario use – e.g., strategic planning for corporations or other organizations, and military and security planning – it is widely understood that there should be close, intensive collaboration between developers and users in the production, revision, and application of scenarios. While high-level decision-makers are not usually involved in the detailed work of scenario construction, they or their surrogates may be intensively involved in problem definition, identification and elaboration of key uncertainties, large-scale scenario design, evaluation and criticism of scenario outputs, and deliberation over lessons and implications. Their level of involvement must be high for results to be useful, particularly if a major purpose of the exercise is to challenge decision-makers' assumptions and promote creative thinking.

In these areas, scenarios typically serve a clearly identified, relatively small and homogeneous set of users who have some degree of agreement on what values they are trying to advance, what issues are relevant, and what choices are feasible, acceptable, and within their power and authority. This is most clearly the case when scenarios are developed for a single organization, but also applies to scenario exercises for larger groups that are sufficiently homogeneous in their interests and perspectives, e.g., scenarios for property and casualty insurers, for organized labor in the United States, or for European environmental groups. In such context, the problems of deciding participation are likely to be manageable.

Intensive user involvement has also been advocated in developing scenarios for climate change. This is obviously correct when climate-change scenario exercises serve specific, clearly identified user groups. The strongest examples are scenarios to support narrowly targeted assessments of impacts and adaptation in particular industries, resources, or regions, e.g., scenarios for coastal managers considering the establishment or revision of setback lines for coastal-zone construction as sea level rises,[170] for rangeland managers considering the purchase of conservation lands or easements for the purpose of providing migration corridors, or for insurance and reinsurance firms examining the nature of climate-change risks they may face and potential responses. In such cases, intensive participation of users is relatively easy to achieve and provides access to valuable expertise and assurance of practical utility.

> Managing a scenario exercise includes deciding who participates, what jobs they are assigned and how these jobs fit together, how disagreements are resolved, and how much time and money are devoted to the exercise. These matters can be decisive for the success of an exercise.

[169] Chess and Purcell 1999; Gregory and McDaniels 2005; Holling 1978; NRC 1996, 2005; Renn et al. 1995.

[170] McLean et al. 2001.

But climate-change scenarios typically serve larger and more diverse sets of users and stakeholders. This is especially true for scenarios produced in large-scale, official assessments such as the IPCC or US National Assessment. Climate-change stakeholders – defined by the CCSP as "individuals or groups whose interests (financial, cultural, value-based, or other) are affected by climate variability, climate change, or options for adapting to or mitigating these phenomena"[171] – are an enormous group, diverse in their interests and responsibilities.

Even when the set of all potential users is numerous and diverse, there may be some types of users who are clearly identified – e.g., climate modelers who need input from emissions scenarios or impact assessors who need input from climate-change scenarios – and who have highly specific scenario needs, including such prosaic factors as data format and resolution. Close consultation with such users is clearly important, especially when their desires exceed what scenario developers can confidently provide, e.g., when climate modelers need emissions data at fine spatial resolution and for specific gases or aerosols, which are not readily available from the energy-economic models used for emissions scenarios. These situations call for particularly close and sustained consultation, so the two sides can understand each other's needs and capabilities in enough detail to develop workable resolutions.

Other users, however, may be numerous, diverse in their disciplinary foundations, methods, and tools, and not clearly identified. Their information needs may have some commonalities but substantial differences. They may even have points of conflicting interest in the construction and use of scenarios. The general case for stakeholder involvement remains strong with

such diverse users, especially in the initial design of a scenario exercise, and in the evaluation and refinement of scenarios for relevance, practicality, and utility. In principle, the required approach is to involve a reasonably diverse and representative group of users and stakeholders, as well as an appropriate range of disciplinary and modeling experts, while keeping the size of the scenario team manageable. But the judgments about participation and representation needed to carry out this approach in any particular scenario exercise will be complex and challenging.

Can a scenario process be completely open? In political settings, some insulation from users may be needed to insure consistency across participating models and analyses. Whatever approach to stakeholder participation is adopted, numbers must be kept manageable. Despite recent progress in scenario methods allowing a substantial increase in the number of participants, there are still practical limits. Although requirements for expertise external to the core scenario team increase with scenario complexity, a scenario process is unlikely to work with a hundred people in the room. A few scenario processes have engaged much larger numbers of participants, but these have greatly reduced the complexity of the scenario-creation process by limiting it to specifying inputs to a single interactive model, or have involved large numbers of participants in independent, parallel sessions interacting with a computer-based model or scenario construction system.[172] These tensions between representational realism, participation, and managerial feasibility pose challenges for design of processes of representation and consultation in scenario development, on which further progress is needed.

Climate-change stakeholders are an enormous group, diverse in their interests and responsibilities.

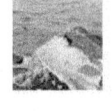

[171] CCSP 2003:112.

[172] See, e.g., Envision Sustainability Tools 1999, Rothman et al. 2003, Stockholm Environment Institute 1999.

> **BOX 4.7. Scenarios in Acid-Rain Assessments: Two Approaches**
>
> Two programs, one in the United States and one in Europe, developed scenarios in integrated-assessment models of acid rain to inform policy decisions over sulfur emissions. Among many other differences, the two programs differed strongly in their approaches to involving stakeholders and in their effectiveness at informing decision-making.
>
> The US National Acid Precipitation Assessment Program (NAPAP) was created in 1980 as a 10-year program to study all aspects of acid deposition: emissions, transport and deposition, impacts, and control strategies.[173] Managed by a committee of six government agencies and supported by a full-time staff office, the program involved roughly 2,000 researchers.[174] Although charged to conduct both scientific research and assessment, NAPAP strongly emphasized research. Its assessment report was opaque on the origin and interpretation of its scenarios, and did not use the scenarios to integrate across the issue or examine implications of alternative policies. Overall, NAPAP is regarded as having succeeded as a research program, but fallen critically short in providing useful information for decision-making.[175]
>
> An alternative approach to acid-rain assessment was taken in Europe as part of the policy debates under the Convention on Long-Range Transboundary Air Pollution (LRTAP). The core of this assessment was a cooperative program to monitor and model acid emissions, transport, deposition, and impacts. In contrast to NAPAP, this program focused more on assessment than research, being specifically established to inform the policy process.[176] Scientific models of components of the acid-rain issue were chosen to contribute to a simplified integration of the problem; scenarios of emissions and controls were chosen in consultation with officials, in an attempt to replicate the policy alternatives under consideration.
>
> The culmination of this pursuit of simple, accessible, and policy-relevant analyses was the RAINS model, developed at the International Institute for Applied Systems Analysis (IIASA) in Austria. As a result of its flexibility, ease of use, and relevance to policies under consideration, the RAINS model was used extensively by policy-makers in the negotiation of sulfur-control agreements under the Convention, and had substantial influence over the controls adopted.[177]
>
> The contrast in approach and outcome between these two programs suggests the potential value of close interaction between experts and stakeholders in producing scenarios, at least when the stakeholders are relatively expert officials responsible for a specific set of decisions. In the European case, such close interaction helped to ensure the credibility of baseline emissions scenarios and the relevance of proposed control scenarios, despite the diverse and sometimes contending interests of the participating officials. The contrast between the two programs also suggests that there can be significant tradeoffs between scientific and assessment objectives in programs that seek to integrate the two activities.

4.4. COMMUNICATION OF SCENARIOS

Scenarios related to climate change must be communicated to multiple audiences, with diverse interests and information needs. Involving users in scenario development can aid subsequent communication in various ways -- e.g., by ensuring that scenarios are understandable and practically oriented, and helping to disseminating scenarios to their constituencies.

But, in all likelihood, most users to whom scenarios must be communicated did not participate directly in scenario development.

Although specific needs will vary from case to case, any communication of scenario-based information to a large, diverse public audience is likely to require a few common elements. First, in addition to the scenarios' content, information should be provided about the process and reasoning by which the scenarios were developed. This allows users and stakeholders to un-

---

[173] NAPAP 1982, Herrick 2002.

[174] Herrick 2002.

[175] Roberts 1991, Cowling 1992, Russell 1992, Miller 1990, Perhac 1991, Rubin 1991.

---

[176] Gough et al. 1998.

[177] Levy 1995.

derstand and critique scenarios, and to determine their own levels of confidence. Second, scenario developers should identify the uncertainties considered. A particularly important distinction to communicate clearly is between scientific uncertainty and scenario uncertainty, including explicit statements of when and how scenarios change (e.g., the reduced SO_2 projections in the IPCC SRES scenarios), and clear explanations of the effects of such changes. Third, related to uncertainty, developers should acknowledge the unavoidable elements of subjective judgment in developing scenarios, and be prepared to explain and defend the judgments they made. Fourth, when particular scenarios were constructed to have specific meanings – e.g., a reference case, a plausible worst-case, or the exploration of a particular causal process taken to its extreme – these should be clearly conveyed. Fifth, if scenario developers have articulated any indicators of the confidence they place on scenarios or distributions of associated variables, this information and any supporting reasoning should also be made available.

A communication strategy should attempt to steer users away from certain common pitfalls, such as choosing one scenario and treating it as a highly confident prediction, or taking the range spanned by a set of scenarios as encompassing all that can possibly happen. An effective strategy of communicating scenarios and their underlying reasoning can help to engage users in the process of updating and improving scenarios. Providing transparency rather than claiming authoritative status for scenarios is likely to increase users' confidence that the scenarios have reasonably represented current knowledge and key uncertainties. It also provides users with the tools to develop alternative representations if they are unconvinced.

In large and complex assessments such as the IPCC and US National Assessment, communication of scenarios and underlying information both to various groups within the assessment and to potential outside users poses representational and managerial challenges. Scenario developers have experimented with various visual techniques for conveying complex information in vivid and understandable form, including landscape representations, maps, and pictures,

as well as various graphical and tabular formats.[178] In the US National Assessment, climate scenarios and other related information were provided to participating assessment teams in several formats (e.g., tabular summaries, models, graphic representations), through websites backed up with workshop presentations. In the IPCC, the Task Group on Data and Scenario Support for Impact and Climate Analysis (TGICA) was established in 1997 to facilitate distribution of climate scenario data, model results, and baseline and scenario information on other environmental and socio-economic conditions, for use in climate impact and adaptation assessments. Data, scenarios, and supporting information are distributed over the internet by the IPCC Data Distribution Center (DDC).[179]

To compactly communicate uncertainty in climate scenarios, the TGICA and several national scenario efforts have developed various graphical methods, including scattergrams showing the range of projected temperature and precipitation changes generated by several climate models using four SRES marker scenarios, and comparing these projected changes to estimates of natural variability.[180] In Figure 4.2, each data point represents one climate-model projection associated with a given SRES emissions scenario. Efforts to develop similarly compact representations of the distribution of scenarios for extremes as well as annual and seasonal averages are underway.

To help users select climate scenarios for impact assessments, an alternative to summarizing climate-model scenarios in such scattergrams is to combine various climate-model results using statistical methods to construct explicit probability distributions for important climate variables.[181] Figure 4.3 shows one such method, which assigns weights to model results

Developers should acknowledge the unavoidable elements of subjective judgment in developing scenarios, and be prepared to explain and defend the judgments they made.

[178] See, e.g., Svedin and Aniansson 1987.

[179] Information on the TGICA is at ipcc-wg1.ucar.edu/wg1/wg1_tgica.html. The DDC is jointly operated by the UK Climatic Research Unit and the Deutsches Klimarechenzentrum, with several mirror sites around the world. Data are provided via the web or CD-ROM. All data distributed are in the public domain.

[180] Ruosteenoja et al. 2003.

[181] Raisanen and Palmer 2001; Tebaldi et al. 2004, 2005.

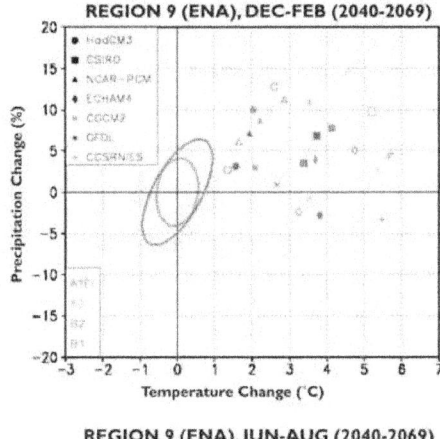

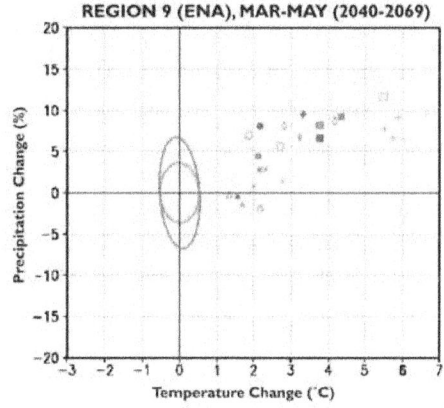

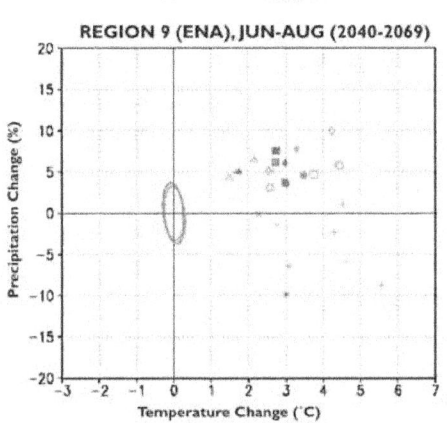

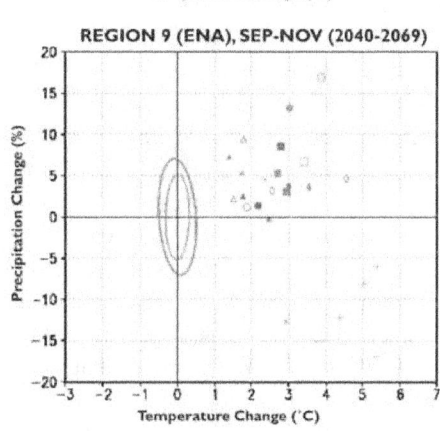

Figure 4.2. Regional scattergram for eastern North America, 2040-2069.

The x-axis shows temperature changes in °C, the y-axis precipitation changes in percent. Each point shows one model's projection under one emissions scenario. A point's color denotes the corresponding emissions scenario, its shape the corresponding model (per legends in upper left figure). Ovals show 95 percent confidence bounds for natural 30-year climate variability, calculated from unforced 1000-year runs of the models CGCM2 (orange) and HadCM3 (blue). Points outside the ellipses indicated projected climate change significantly outside the range of natural variability, most frequently due to changes in temperature rather than precipitation. *(Source: Ruosteenoja et al. 2003.)*

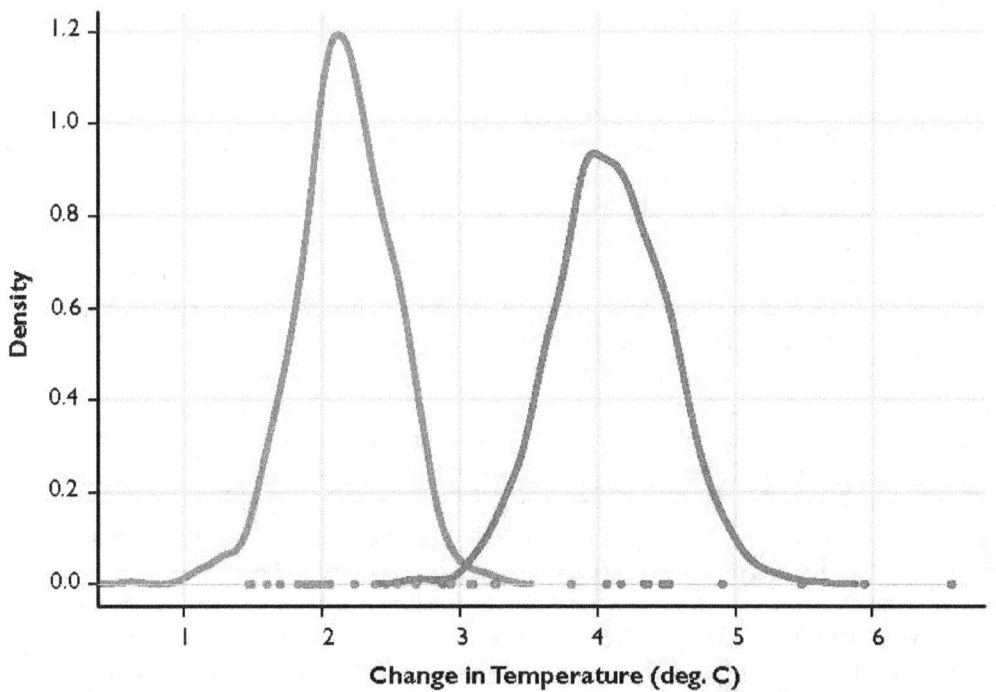

Figure 4.3. Constructed probability distributions of model-simulated temperature change in 2080-2099

The x axis shows projected temperature change in Eastern North America from the 1980-1999 historical average, using 19 climate models participating in the IPCC Fourth Assessment Report driven by the SRES A2 (red) and B1 (blue) emissions scenarios. Each point on the x axis shows the result from one model. The curves above the axis show probability distributions constructed from these individual model results. *(Source: Tebaldi et al. 2005.)*

based on their bias in simulating the current climate (smaller biases are assigned higher weight) and their correspondence with other model results (outliers are assigned lower weights). This method compactly communicates multiple model results, clearly conveying which ones fall at the top and bottom of the distribution ("unlikely to be higher/lower than this"), and which fall in the middle of the range.

This current focus on collections and intercomparisons of model-based projections with various emission scenarios represents a new approach for communicating scenario-driven model output to users engaged in assessment and adaptation activities. It has enabled users to consider a broader range of emission scenarios and climate models than was feasible in the US National Assessment and previous IPCC assessments. It allows users to consider all available model and scenario combinations to span the literature, or to select only scenarios that exceed some threshold of interest or fall within some specified probability range. Future assessments should benefit from this type of multi-model, multi-scenario approach, which allows users more effective and informed choice over scenarios to consider.

4.5. CONSISTENCY AND INTEGRATION IN SCENARIOS

One of the most often stated requirements for scenarios is that they be "coherent" or "internally consistent." This is clearly an important goal. Scenarios usually specify multiple characteristics of an assumed future, whether as multiple elements of a narrative or multiple quantitative variables, so these elements should fit together. Difficulties arise in the pursuit of such consistency, however, and in some scenario exercises the pursuit of consistency, together with the goal that scenarios integrate many components of a broad issue such as climate change, may jeopardize the validity and usefulness of the scenarios.

Certain elements of internal consistency in scenarios are unproblematic, such as avoiding gross contradictions with well known principles of behavior of biophysical or socio-economic systems, and not inadvertently moving far outside the bounds of historical experience. Inad-

vertently implausible assumptions can arise, for example, when multiple elements of a scenario are specified without cross-checking; e.g., end-year specifications of a region's population and GDP without checking the implied growth rate in GDP per capita, or specifying energy-related emissions trajectories without checking what they imply for resource availability. Avoiding these pitfalls requires thorough cross-comparisons of related values with each other, of terminal values with implied time-trends in the intervening period, and of values within and among regions. Scenario developers should not always and necessarily avoid extreme or unprecedented outcomes, however. Presenting extreme or seemingly implausible future conditions *intentionally*, with an explanation of how they could in fact arise, can contribute to several of the major purposes of scenarios, e.g., shaking up habitual thinking and broadening expectations of what future developments are plausible.

But statements about internal consistency in scenarios usually claim much more than the mere absence of gross contradictions and inadvertently implausible values. They tend to claim that the multiple elements of a scenario are related in a way that reflects reasonable, well-informed judgments about causal relations, suggesting that some events or trends are more likely to occur together, some less. Expressing the goal as "coherence" rather than "internal consistency" suggests a higher level of perceived affinity among scenario elements, evoking normative or even aesthetic aspects.

Expressed in probabilistic terms, statements about internal consistency may be interpreted as claims that alignments of factors similar to those in the scenario are more likely than other, dissimilar alignments. One might, for example, claim that a scenario with rapid growth in both the economy and energy use is more internally consistent than one in which the economy grew rapidly but energy use did not. But where do these perceptions of greater or lesser likelihood come from, and how valid are they? In some cases a well-founded theory or model might say that certain outcomes tend to be related. Alternatively, explicit analyses might connect the claim to underlying assumptions that are open to scrutiny and criticism. But in the absence of

The current focus on collections and intercomparisons of model-based projections with various emissions scenarios represents a new approach for communicating scenario-driven model output to users engaged in assessment and adaptation activities.

such transparent foundations for judgments of what scenario conditions are consistent and what are not, these claims can only rest on more diffuse judgments by scenario developers, refined and tested through various deliberative processes – e.g., arguing about the claims, working through their implications relative to those of alternative specifications, and identifying additional bodies of research and scholarship that can be brought to bear.

These difficulties can be compounded when, in addition to consistency, a goal of scenario "integration" is also pursued (although the precise meaning of "integrated" can be difficult to ascertain). The integration of a scenario is a function of its complexity or breadth, which is related to the number of characteristics it jointly specifies. In global climate-change scenario applications, integration typically refers to including all major elements of the causal chain of the issue, i.e., multiple dimensions of emissions and their socio-economic drivers, climate, impacts of climate change, and responses.

Asking a scenario to be integrated in this way imposes on the scenario the burden of capturing all relevant elements of the future. Such an expansive scenario may occasionally be needed – e.g., for preliminary assessment of a threat for which no relevant data or current research exists. However, the risks of error, bias, and arbitrariness in such a scenario are greatly increased, because so much of reality (with whatever unknown causal processes by which it actually operates) is being stuffed into the scenario.

More likely, an integrated scenario would be constructed by combining exogenous assumptions about some elements with model-calculated values for others. This approach does not avoid increasing risks of inconsistency and contradiction as a scenario is expanded, particularly when multiple models are used. Since models embody specific, quantitative causal relations among variables, they do not require – or indeed allow – all variables to be specified. Scenarios provide only those exogenous inputs that the model does not produce. These scenario-based inputs should be consistent with each other, but to a lesser extent than the precise standard that defines consistency in a scenario. These ex-

ogenous inputs, together with model results, can jointly comprise a scenario that is generated for some alternative use.

Consistency problems grow when scenario exercises involve multiple models and attempts are made to achieve model harmonization. When scenarios are constructed partly out of exogenous inputs provided by a scenario (made consistent as much as possible through qualitative or intuitive causal reasoning) and partly out of models, multiple models are often used. Using multiple models in parallel can allow for more extensive exploration of causal relations, and helps to characterize uncertainty in scenarios since different models embody different representations of causal processes. It may also enhance the credibility of the process. But models of the same broad set of phenomena – e.g., models of the economy and energy sector – frequently differ in which variables they require as exogenous inputs and which ones they calculate endogenously. In this case, some variables must be specified exogenously for some models, but are calculated endogenously by others.

This creates various problems for consistency. In general, when scenario exercises are conducted in this way, some elements are assumed and others are model-calculated. Attempting to avoid this poses even more serious problems, however. It is not usually possible to arbitrarily perturb the exogenous input variables so all inputs and outputs match across all models, since such perturbations will influence other variables in the model. Consequently, avoiding these inconsistencies will require manipulating internal relationships within models to make their outputs match the specified values, given the common inputs. But such reverse-engineering of internal model relationships to match specified outputs, in addition to being exceedingly cumbersome and arbitrary, can corrupt the internal logic of models, obscure the interpretation and significance of results, and make it impossible to use model variation to illuminate uncertainty.

For example, in an exercise to generate non-intervention scenarios of potential future emissions, little insight is likely to be gained from defining scenarios in terms of the resulting emissions and forcing the different models to

In the absence of transparent foundations for judgments of what scenario conditions are consistent and what are not, claims can only rest on more diffuse judgments by scenario developers, refined and tested through various deliberative processes.

generate these emissions targets.[182] It may be equally fruitless to define scenarios in terms of GDP and energy consumption trajectories and to force multiple models to reproduce these. For this reason, multi-model exercises such as the Energy Modeling Forum usually seek to harmonize only a few of the most essential and commonly used inputs.[183] When multiple models are used to generate scenarios, the most useful way to pursue consistency may be to develop common assumptions for the variables furthest back in the causal chain, but the wide variation of model structure can make even this approach to harmonization challenging.

In addition to consistency within a scenario, consistency across scenarios within an exercise also requires attention. Ideally, factors not explicitly recognized as the basis for inter-scenario differences should be consistent across scenarios. Or alternatively, all bases for differences between scenarios should be explicitly recognized and stated.

When models are used in a scenario exercise, significant variation in model structures suggests less mature underlying knowledge, or at least greater recognition of knowledge gaps, than when model structures converge and all remaining uncertainty is over exogenous input parameters. For scenarios to provide faithful representation of present knowledge and uncertainty, this variation should not be suppressed or concealed. Consequently, when scenarios are defined over variables that include outputs of some participating models as well as inputs, it is crucial not to pursue false consistency by forcing models to match the target outputs through manipulation of their internal causal processes. This is suppressing model uncertainty.

One preferable alternative would be for the results of scenario exercises involving both exogenous inputs and multiple models to

> When multiple models are used to generate scenarios, the most useful way to pursue consistency may be to develop common assumptions for the variables furthest back in the causal chain, but the wide variation of model structure can make even this approach to harmonization challenging.

explicitly distinguish between three classes of variables: (1) a minimal set, exogenous to all; (2) those specified exogenously for some models, but generated by others; (3) model outputs, whose variation reflects partly model and partly parameter uncertainty.

An alternative way to use multiple models is to let each model produce one scenario, as was done in the selection of the SRES marker scenarios. With this approach, each scenario represents a particular realization of uncertainty over both exogenous inputs and model structure. But this approach confounds model uncertainty with parameter uncertainty. It may be preferable to cross exogenous inputs with models to produce a larger number of scenarios from which subsets can be extracted as needed, perhaps organizing these as a nested hierarchy of scenarios similar to the SRES: six marker scenarios, 40 SRES scenarios in total, and hundreds of scenarios in the literature review.

There are good reasons to combine narrative with quantitative approaches, as scenario exercises have increasingly sought to do. But the connection between qualitative and quantitative aspects of global-change scenarios has been inadequate, diminishing the usefulness of the exercises due to inconsistencies within each type of scenario and between the two types. This problem has partly been due to limited time and resources, but has also reflected substantive difficulties in linking the two types of scenario, difficulties that have not been understood or managed well. Narrative scenarios typically specify deep structural characteristics like social values and the nature of institutions, which are associated with structural characteristics of models such as the determinants of fertility trends, labor-force participation, savings and investment decisions, and substitutability in the economy. Consequently, the differences among alternative narrative scenarios, reflecting different basic assumptions about how the world works, correspond more closely to variation of model structure than to variation of parameters. Better integrating the two approaches will require developing ways to connect narrative scenarios to model structures, rather than merely to target values for a few variables that models are then asked to reproduce. This has not hap-

---

[182] Note that this is not the case if the purpose of scenarios is to explore the implications of specified limits on future emissions. If an emission constraint is assumed to be imposed by policy, then different models can be used to explore the implications of that constraint for costs, technologies, and other impacts. In this case, caution is needed in deciding what other model variables, if any, should be constrained.

[183] Weyant and Hill 1999.

pened because scenario exercises have not had the capability or resources to direct new model development, or to induce modelers to undertake substantial structural changes to their models. This would require substantial efforts, including getting modelers to interact with scenario exercises in a new way, but might hold more promise for allowing scenarios to usefully inform discussions about large-scale policy choices for mitigation and adaptation.

4.6. TREATMENT OF UNCERTAINTY IN SCENARIOS

Representing and communicating uncertainty is perhaps the most fundamental purpose of scenarios. This section discusses how scenarios represent uncertainties, how these methods connect scenario exercises to simpler formal exercises in the analysis of decisions under uncertainty, and what challenges are posed in how uncertainty is represented. It also addresses several important debates in the treatment of uncertainties.

In most scenario exercises, uncertainty is represented not in a single scenario, but in variation across several scenarios considered together.[184] The choices to be made in deciding how to represent uncertainty include the following:

- What characteristics are varied

- By how much these characteristics are varied, separately and together (e.g., should extreme values of multiple characteristics be combined, or extremes of some combined with the middle cases of others?)

- How many scenarios to create and consider together

- What description, documentation, or other information is attached – including whether and how specifically measures of likelihood are assigned.

4.6.1. Uncertainty in simple quantitative projections: basic approaches

How these choices are made and their implications for scenario use and effectiveness are closely related to the large-scale decisions in designing a scenario exercise outlined in Section 2.1. In particular, the role of uncertainty in a scenario exercise is strongly linked to scenario complexity, richness, and use. In the simplest case, a scenario exercise may be dominated by a single quantitative variable, so all uncertainty could be represented by alternative future levels or time-paths of that variable. This case is so simple that many scholars and practitioners argue it should not be considered a scenario at all.[185] Still, even this simple and extreme case raises significant issues. We begin here and then move to more complex cases.

If we also assume that the probability distribution is known, the situation reduces to a formal exercise in analysis of decision-making under uncertainty. Given a known set of choices and outcomes of each choice under each uncertain outcome, alternative choices can be evaluated by formal methods such as seeking the best outcome on average or under some risk-averse valuation scheme, or seeking robust strategies. This decision-analytic approach can be extended to situations of a few uncertain variables with a known joint distribution, multiple decision-makers who evaluate outcomes differently, or (with somewhat more difficulty) decision makers with different probability distributions.

Further relaxation of these simplifying assumptions moves us toward activities that are more widely recognized as scenario exercises. First, if a scenario exercise is addressed to more than just a few decision-makers with known choice sets and outcome valuations, scenarios can no longer simply be inputs to an analytic exercise, but rather become descriptions of potential future states that must be communicated directly or indirectly to decision-makers for their reflection and deliberation. Second, if distributions of important quantities are unknown, it is necessary to exercise judgment regarding how to

Representing and communicating uncertainty is perhaps the most fundamental purpose of scenarios.

---

[184] When a scenario exercise uses just one scenario, this usually presents some specific threat or challenge posed to existing procedures or decision-makers. In these cases, uncertainty is still represented by differences among scenarios, but the single scenario is implicitly contrasted to the status quo.

[185] E.g., Wack (1985a: 74) states that such a scenario is just "quantification of a clearly recognized uncertainty."

draw on relevant knowledge to construct and describe alternative future values of the quantities, and how to represent these values to users with a manageable number of scenarios.

Since scenarios describe future conditions, the distributions of quantities in scenarios cannot be known in the same sense that the distribution of current characteristics – e.g., the November daily high temperature at O'Hare Airport – can be known through repeated observations. Probabilistic statements about future conditions always incorporate elements of subjective judgment. Many forms of current knowledge – including data, models, and expert judgments – are relevant to forming these judgments about future conditions. In constructing scenarios of population growth, for example, the distribution of observed past growth rates can be used to construct a range or distribution of plausible future values. But while scenarios can draw on, and be made conditional on, such knowledge, this does not overcome their unavoidable reliance on subjective judgments as well.

Scenarios can also be based on model representations of knowledge of causal processes. For example, instead of simply extrapolating past population growth rates, one could use a demographic model that represents trends in fertility rates, lifespan, and migration to calculate a resultant population trend. Formal modeling can represent the structural relationships transparently, reducing the risk of generating inconsistent projections. Structural models can possibly also perform better in extrapolating to conditions beyond the observed range of behavior. Because models represent causal relationships among multiple variables, these models can extend the range of current and historical data that are relevant to projections, although this may result in an expansion of data needs. Models can also help characterize uncertainty in future quantities of interest, by allowing the uncertainty to be attributed to input parameters – explored through sensitivity analysis or simulation techniques such as Monte Carlo – or to model structure.

Estimating output distributions based on assumed distributions of uncertain input parameters does not capture all uncertainty of importance for

assessment and decision-making. The input probability distributions are not known with certainty, nor are the structural assumptions that determine the mapping of inputs onto outputs within any particular model. Uncertainty analysis can embrace this additional level of uncertainty, sometimes called "meta-uncertainty," by stepping up one more level of abstraction – considering not just uncertain quantities, but uncertainty about their uncertainty, or alternatively, probability distributions over probability distributions of unknown quantities. Methods to represent and process such meta-uncertainty mirror those used for first-order uncertainty. This is an active area of research, but its importance for assessment methods and their application is unclear. This level of abstraction increases the difficulty of communicating scenarios and their underlying reasoning transparently and comprehensibly to non-specialists. Moreover, since any step of analysis represents an act of potentially fallible judgment, taking the step to meta-uncertainty still does not capture all possible uncertainty. It is not clear whether, for purposes of constructing and using scenarios, the explicit separation of uncertainty in outcomes from uncertainty in probability distributions brings more benefit than could be gained from simple heuristic guidance to assume distributions are wider than initially seems necessary.

A major risk in all scenarios is subjective bias, which can be reduced but not eliminated through use of existing data and formal modeling. Judgment is an essential element in constructing scenarios, both to apply relevant data and models when these are available, and to build future descriptions using less formal methods when they are not. The expert judgments supporting such less formal projections may be better founded than mere uninformed speculation, since there is typically much relevant knowledge available beyond what is explicitly captured in present datasets and models.

Approaches to developing expert-judgment based projections vary widely in their structure and formality, from simply asking one or more experts to state their best estimate of some unknown quantity, to highly structured elicitation exercises that provide multiple cross-checked

Probabilistic statements about future conditions always incorporate elements of subjective judgment. Many forms of current knowledge – including data, models, and expert judgments – are relevant to forming these judgments about future conditions.

estimates of the same quantity.[186] Such methods must attend to risks of overconfidence and bias, which are well documented in experts as well as laypeople. Carefully designed elicitation protocols can reduce the effects of such biases, e.g., by prompting experts to broaden their estimates of uncertain quantities, but cannot eliminate them.[187] An additional challenge to these methods is that there is no generally accepted method for selecting or aggregating estimates from multiple experts.

4.6.2. How many scenarios, over what range?

In communications of scenarios, limited time, resources, and attention usually require that only a few discrete values or time-paths are specified, not a complete distribution. Scenario developers must decide how many scenarios to provide and how to space them.

How many scenarios to provide rests on a judgment of the value provided by each additional point from the underlying distribution relative to the burden of producing and using each new scenario. If the use made of each scenario is expensive – e.g., consuming large quantities of time of busy senior people, or running a large model – then the number of scenarios that can be adequately treated may be very few. The 1992 IPCC scenario exercise provided six separate scenarios, of which nearly all subsequent analyses used just one or two. Of the 40 scenarios produced by the SRES process, only 6 (initially 4) were highlighted as "marker" scenarios, while most subsequent analyses used just 2 or 3.[188]

Deciding how many scenarios to provide also involves some element of attempting to avoid predictable errors in their use. While the most obvious and frequent choice in providing scenarios of a quantitative variable has been to provide three – one high, one low, and one in the middle – it has been widely noted that this practice runs the risk that users will ignore the top and bottom, pick the middle, and treat it as a

highly confident projection, suppressing the uncertainty that scenario developers tried to communicate by providing, and carefully spacing, three scenarios. The same risk applies to any odd number of scenarios, leading many developers of quantitative scenarios to the informal guideline that the number provided should always be even, so there is no "middle" scenario that users can inappropriately fix on.

More specific guidance on the appropriate number and range of scenarios must reflect both scenario developers' sense of the underlying distribution from which scenarios are drawn, and their intended use. One must consider whether departures in both directions from the middle are of similar importance, or whether only departures in one direction need be represented. For example, one might judge that in an assessment of impacts of climate change a scenario drawn from the lower tail of potential climate change is likely to provide little substantive insight, since in most cases the impacts of a small-change scenario are predictably small.

One must also consider how far a set of scenarios should extend toward including extreme or unlikely futures. In estimating unknown quantities, many fields of empirical research draw intervals to capture from 90 percent to 99 percent probability, but in constructing scenarios to inform decisions there may be good reasons to consider more extreme and less likely possibilities, whether these likelihood judgments are expressed quantitatively or qualitatively. Assessments and policies in both regulation of health and safety risks and national security, for example, routinely focus on high-consequence risks that are judged much less than 1 percent likely. Similarly for global environmental change, low-probability risks might need to be considered if their consequences or their effects on preferred decisions are large enough.

It is often suggested that a set of scenarios should "span the literature" of prior scenarios or projections of the same quantities. However, there may be good reasons for a wider or different range, or even a narrower range – although developers should be cautious about a set of scenarios that spans a much narrower range than published estimates of the same quantities. A published scenario may have been

Similarly for global environmental change, low-probability risks might need to be considered if their consequences or their effects on preferred decisions are large enough.

---

[186] Morgan and Keith 1995.

[187] Tversky and Kahnemann 1974, Wallsten and Whitfield 1986.

[188] Initially A2 and B2 were most widely used. More recent work has used A2 and B1, sometimes with A1B.

constructed to serve various purposes other than providing an independent new estimate of a quantity of interest. Previous scenarios developed to serve some particular purpose may or may not be relevant to a new scenario exercise, depending on the relationship between their intended purposes. Moreover, previously published scenarios can be highly self-referential, since many published analyses use prominent pre-existing scenarios as inputs to a new study, or examine a new model by forcing it to reproduce some pre-existing scenario. For all these reasons, previously published scenarios are better regarded as one input to the judgment of developers of new scenarios than an authoritative picture of present knowledge that new scenarios must follow.

4.6.3. Bifurcations and major state changes

While many uncertainties may be treated as a continuous range of possible values, some uncertainties may capture large-scale bifurcations or abrupt changes. For climate change, potential abrupt changes include melting of major continental ice sheets or shifts to some new mode of ocean circulation.[189] Large-scale bifurcations may also arise from breakthroughs in energy technology. Such possibilities are typically not captured in either historical data (because they are by assumption novel), or models (because they would represent a change in the causal structure represented in models).

Abrupt changes can pose particular challenges for deciding the number and range of scenarios to include in an assessment or decision-support exercise, either because their consequences are so extreme or because they would fundamentally change our understanding of how the system operates. The decision of whether and how to consider these uncertainties consequently turns on the balance between their probability – which is believed to be low but not well characterized – and their high consequences, which must be evaluated relative to the scenarios' intended use. This will be a particularly difficult choice when only a few scenarios are being generated. For example, in a coastal impacts assessment the enormous consequences of the

difference between a half-meter and a five-meter sea level rise over this century – and the well-identified mechanism by which such a large rise could occur – may suggest the importance of explicitly considering a scenario involving loss of one of the major continental ice masses. But including such a scenario runs the risk that users will assign it a much higher probability than is appropriate, either because of its vividness and extremity or because they presume that developers' decision to include the scenario meant that they assigned high probability to it. When such a scenario is included, scenario developers have a serious responsibility to communicate, loudly and consistently, its different status.

A further challenge in representing large-scale or discrete changes in scenarios is that there might be many such possibilities, all of them high-consequence but believed to be unlikely. Including any particular one may mislead both by exaggerating its likelihood and by strengthening users' tendency to ignore others, when these all represent "unknown unknowns" that should receive some consideration. The more there are, the more the right approach might be to shift all scenarios further out to reflect the various mechanisms by which conventional understanding may under-represent the tail of the distribution, rather than highlighting any particular abrupt-change mechanism by giving it a scenario of its own.

4.6.4. Uncertainty in multivariate or qualitative scenarios

As the characterization of future conditions within scenarios grows more complex, so does the process of representing uncertainty within them. While many of the issues discussed above in the simplified context of scenarios on a single variable also apply to multi-dimensional scenarios, several additional issues arise.

The most basic of these is that with multiple dimensions of variation in scenarios, it is necessary to decide which uncertainties are represented. Even when scenarios include only multiple quantitative variables, it is no longer possible for a few scenarios to span all corners of the joint distribution of these variables. Rather, they must combine variations in ways

Previously published scenarios are better regarded as one input to the judgment of developers of new scenarios than an authoritative picture of present knowledge that new scenarios must follow.

---

[189] NRC 2002.

that are most illuminating and important for the purpose at hand, massively reducing the dimensionality of the problem to make it intelligible for users. In addition, increasingly detailed and realistic scenarios often specify characteristics that are qualitative, or described less precisely than cardinal variables. For example, alternative scenarios might specify that current trends of globalization increase, stagnate, or reverse, or that decision-making capacity on climate change increases or decreases. Such characteristics may be judged crucial to include because they may be among the most important drivers of preferred choices or consequences of concern.

Scenarios of this kind pose substantial further challenges in representing uncertainty and interpreting its meaning. Relative to the simple quantitative scenarios we have considered up to this point, these lie in a much higher dimensionality space of future possibilities; they may not lie in any ordinal relationship to each other; and they may include characteristics whose definitional boundaries are not precisely specified. Defining a small set of scenarios to reasonably span the most important uncertainties is consequently even more difficult than for simple quantitative scenarios.

The approach most widely proposed to represent key uncertainties in such scenarios is to seek underlying structural uncertainties that satisfy two conditions: they appear to be most important in influencing outcomes of concern or relevant decisions; and they are linked with variation in many other factors. These underlying uncertainties can be simple discrete states such as peace or war, prosperity or stagnation; or, as in several major global environmental scenarios, they can be deeper societal trends, such as more or less globalization or shifts in societal values toward greater environmental concern, from which variation in many factors is assumed to follow.

This approach, formalized in the Shell scenarios method,[190] involves two steps: first identifying a small number of fundamental uncertainties and a small set of alternative realizations of each; and then elaborating additional future charac-

teristics associated with each realization through both qualitative reasoning to fill in a narrative, and assembly of data and model results to build a parallel quantitative description to the extent this is judged useful. Repeated, critical iteration between the qualitative and quantitative elements is conducted, to bring additional relevant knowledge and expertise to bear and to check for consistency.

Even rich narrative multivariate scenarios must imply certain claims of likelihood. Every scenario included must be deemed likely enough to merit the resources and attention spent on developing and analyzing it. This applies even to extreme-event scenarios that are intentionally constructed to capture the low-probability tail of the distribution, since even they must be perceived likely enough to merit time and attention given their severity. Since users would reject any scenario that they persistently judged too implausible to consider, when decision-makers find a scenario exercise useful, it validates developers' judgment that each scenario was likely enough to consider.

In a purely mathematical sense, any one specific rich multivariate scenario must be arbitrary and of vanishingly small probability. There are, however, ways in which it may be reasonable to assign non-zero probabilities to multivariate scenarios. First, if scenario designers in fact succeed at identifying a few deep structural uncertainties that strongly condition outcomes on many other characteristics in a scenario, then the richness of a scenario description need not imply that it is vanishingly unlikely. Whether this is true or not is a judgment to be made by scenario developers and users in each application. If they are sufficiently careful in their development and critical examination of scenarios, their judgment may well be correct. But there will often be no way to further test these judgments, so it is of course possible that the proliferation of additional detail in scenarios – even detail that developers and users recognize is crucial for determining valued outcomes and preferred choices – is arbitrary or erroneous.

A second way in which rich, detailed scenarios may be judged sufficiently likely to consider concerns the precision with which scenario characteristics are specified. In rich multivari-

The approach most widely proposed to represent key uncertainties in scenarios is to seek underlying structural uncertainties that satisfy two conditions: they appear to be most important in influencing outcomes of concern, and they are linked with variation in many other factors.

---
190 Shell International 2003.

ate scenarios, many characteristics are often specified diffusely: economic growth may be merely "high" or "low," rather than being stated as a particular value. Even when a characteristic is stated quantitatively, its particular value may be treated as merely illustrative of a range of similar values; e.g., annual GDP growth might be set at 4 percent because a user needs a numerical model input, but it is understood to represent a broad range of similar values that all count as "high" growth. Interpreted in this way, a multivariate description may remain likely enough to merit examination – and indeed, a modest number of scenarios may exhaust the set of potential futures that matter for the issue at hand. Here one is not assigning likelihood to the precise numerical assumptions used to flesh out the details of a scenario, but rather to a thick slice of future conditions that resemble that scenario more than the other scenarios in the set.

4.6.5 The debate over quantifying probabilities

A major debate in the use of global-change scenarios has concerned whether or not to specify quantitative probabilities associated with scenarios. This debate is central to the meaning and use of scenarios, and has been sharpest over the IPCC's SRES scenarios. Developers of the SRES scenarios decided at the outset of their process that they would make no attempt to assign probabilities to scenarios, in part because they were adopting the Shell approach of developing scenarios from storylines, in which quantitative probabilities are usually avoided. After the scenarios were published, several critics argued that since the most prominent and important outputs were the projections of emissions under the six marker scenarios, it was natural – and essential for development of rational climate-change policy – to describe the distribution of emissions in probabilistic terms. For example, how likely are 2100 emissions to lie above the 30 GtC of scenario A2 or below the 5.2 GtC of B1? Should the range spanned by all 40 SRES scenarios be understood to comprise 90 percent of all probability? 99 percent? All of it?

Developers of the SRES scenarios stood by their initial decision not to quantify probabilities. Since the controversy only became promi-

nent long after the decision had been made by a writing team no longer in operation, it would have been virtually impossible for the group to retrospectively assign such probabilities. But rather than rely on this argument of managerial infeasibility alone, SRES organizers offered a vigorous substantive defense of their initial decision. This defense relied in part on the statement that the six marker scenarios were all "equally sound," without providing any guidance regarding what this meant other than explicitly denying that it meant "equally likely." Describing each of the six marker scenarios as "equally sound" represents the entirely reasonable case that in the developers' judgment these all needed to be considered seriously – without making any further judgment as to their likelihood. While clearly frustrating to those wanting to use the scenarios as a basis for policy, the result is entirely consistent with the IPCC mandate to do assessment, but not to reach policy conclusions.

However, this debate will continue; it rests in part on different conceptions of the meaning and typical contents of a scenario. The simpler the contents of scenarios, the more readily they lend themselves to explicit quantification of probabilities. When scenarios consist only of alternative time-paths of a single quantitative variable, or one such variable is of predominant importance, it is straightforward and sensible to understand the intervals between those time-paths to have probabilities associated with them and there are several strong arguments for being explicit about these probabilities. First, stating probabilities allows comparative risk assessment between scenarios and explicit exploration of risk-reducing strategies.[191] Second, sophisticated decision-makers whose choices depend on uncertainty in these variables need probability information about possible values, not just a set of alternative values, to evaluate choices – whether their approach to decision-making is based on expected values, risk-aversion, seeking robust strategies, or some other approach. Finally, when such scenarios are presented without probability judgments, users may attach their own, often via simple heuristic devices that may misrepresent the developers' understanding. For example, many subsequent users of the

> The simpler the contents of scenarios, the more readily they lend themselves to explicit quantification of probabilities.

---
[191] Webster 2003.

SRES emissions scenarios have simply assumed the probabilities they needed to conduct further assessments, using such simple devices as counting scenarios or assuming a uniform distribution over the entire range.

Opponents of explicit quantification of probabilities do not dispute that such probabilities can coherently be assigned to simple scenarios in one or two quantitative variables. Rather, they raise principled objections to the appropriateness of attempting to quantify probabilities for more complex scenarios, particularly those involving socio-economic conditions, as well as practical objections to the use of probabilities even in the case of simple quantitative scenarios.

Many researchers are less comfortable using probabilities for complex scenarios that include explicit socio-economic elements than for uncertainties that are purely bio-physical, such as probabilities of different rates of climate change, conditional on a particular emissions scenario. Four main arguments are advanced against the use of probabilities for such scenarios.

First, some argue that the large multivariate space of possibilities from which such scenarios are drawn, and the vague and qualitative way that some scenario characteristics are specified, make it impossible to coherently define the boundaries of the outcome space to which probabilities are being assigned. There is no way to clearly define the interval "between" one scenario and another; and if probability is attributed to a slice of possibilities around each scenario rather than to the intervals between them, is it not possible to define clearly the boundaries of the slice to which the probability is assigned. To the extent that scenarios describe different types of worlds, which are distinguished from each other by alternative resolution of a few key uncertainties – e.g., high or low growth, high or low globalization – where the location of the boundary is not precisely specified, it may be difficult to create a shared understanding of these boundaries between users and creators. But if assigning a precise numerical probability is judged too difficult in these cases, less precise descriptions such as "highly likely," "more likely," "less likely," or "roughly equal" could be assigned. In some applications where scenarios are intended to cap-

ture all the uncertainty of concern to the decision-maker – i.e., scenarios are intended to be mutually exclusive and collectively exhaustive – there may even be a reasonable basis for numerical probability.

The second argument for rejecting probabilistic description of socio-economic conditions is based on "reflexivity" – the proposition that scenarios may influence the behavior or decisions driving the scenarios, so probability judgments about scenarios could reflect back on themselves, becoming either "self-fulfilling" or (more plausibly) "self-denying" prophecies. Section 4.1 addresses this issue in some detail, in particular in the distinction between how to treat mitigation decisions in scenarios to inform mitigation decisions and impacts or adaptation-related decisions. We might only add here that for scenarios of global emissions, reflexivity could only operate if both the influence of scenario judgments on their users' behavior and the influence of their users' behavior on global emissions were extremely strong. Moreover, it is not evident why scenarios with explicit likelihood judgments should raise this concern, while scenarios presented without such judgments – which also presume some claims of plausibility or likelihood – should not. Concern about reflexivity appears more serious for scenarios prepared in close consultation with national mitigation policy-makers, and it is for this reason (among others) that we judge explicit attempts to assign probabilities less valuable for scenarios prepared in such settings.

Third, some argue that it should not be scenario developers or experts who make judgments about likelihood of alternative scenarios, but users – particularly when scenarios are used to inform high-stakes public decisions. But this depends on the details of the content and use of scenarios. For some scenario elements in some settings, particularly use of scenarios to advise specific policy decisions, the scenario users may be as expert as the developers in associated uncertainties and risks, or more so. But in such settings, the use of scenarios normally highlights critical examination of these assumptions, and users have the knowledge and assertiveness to probe, critique, modify, or reject scenario elements that they find weak, including probability judgments. When scenarios are produced to

Concern about reflexivity appears more serious for scenarios prepared in close consultation with national mitigation policy-makers, and it is for this reason (among others) that we judge explicit attempts to assign probabilities less valuable for scenarios prepared in such settings.

Overall, we find the arguments in favor of quantifying probabilities to be strongest for scenarios whose major outputs are projections of one quantitative variable (or very few), and weakest for complex multivariate scenarios with substantial qualitative or narrative elements.

serve many diverse users and consequently cannot rely on intensive interplay with representative, well-informed, and challenging users, scenario developers frequently have the best access to available knowledge relevant to forming probability judgments. Not making these judgments explicit is withholding information that users may need to understand and interpret the scenarios. If scenarios and their underlying reasoning and assumptions are presented clearly enough, users can make informed choices whether or not to use probability judgments that are provided.

Finally, some argue that probabilities cannot be known, or even sensibly estimated, for socio-economic futures – perhaps because socio-economic processes and mechanisms are intrinsically less knowable than biophysical ones, perhaps due to the unpredictable effects of human creativity and leadership, and perhaps because causation does not operate in the human domain as it does in the bio-physical domain. Although these arguments raise deep philosophical questions, as a practical matter probabilistic projections are routinely done in some socio-economic domains, including projections of population and economic growth, but not, or not well, in others, such as projecting technological innovation. Provided the basic concept of subjective probability is accepted, weaker knowledge and deeper uncertainties can be accommodated by broadening the relevant uncertainties rather than declining to make probabilistic judgments, but the question remains of whether the resultant broad uncertainty ranges are meaningful or operationally useful.

Several practical objections have also been raised to associating explicit likelihood judgments with scenarios. These include the difficulty of developing probability estimates from multiple information sources that can achieve sufficient agreement from diverse experts, and the non-intuitive nature of probability distributions in using scenarios to communicate with non-expert users. These are both valid concerns, although active areas of research and development in expert elicitation techniques and in simple intuitive devices to communicate uncertainty are making some progress in mitigating them.

An additional practical argument against quantifying probabilities is that attempting to do so may represent a distraction that uses time, generates conflicts, and is of little value to scenario users. Whether this is indeed the case, however, is in part a judgment to be made by scenario users, not developers. Opponents of quantified probability argue that users typically only need scenarios to pass some probability threshold. Beyond this threshold, they will seek robust choices that yield acceptable outcomes under all possibilities, so further refinement of probability serves no purpose. This argument has merit, but only to the extent that it accurately describes how these scenarios will be used. Quantitative assignment of probabilities to scenarios when high-stakes decisions are implicated is clearly difficult and contentious, as the SRES controversy illustrates. Even if this argument correctly characterizes how scenarios are used, users might still be able to profitably exploit more detailed probability information if it were available – although one must also consider the risk that non-technical users might somehow be more likely to misunderstand scenarios with explicit probability judgments attached (perhaps by taking a stated probability distribution as the "true" distribution) than to misunderstand a simple collection of scenarios presented with no such probability information (perhaps by taking the range presented to embrace the totality of all possibilities). It is also possible that engaging scenario users in an attempt to assign probabilities, even only illustratively, could both draw on relevant knowledge of uncertainties that they possess more than scenario developers, and provide a valuable device to probe and sharpen their understanding of the situation. Any argument based on the information needs of specific users becomes less persuasive as the set of potential uses and users, and the likely diversity of their information needs, grow larger.

Overall, we find the arguments in favor of quantifying probabilities to be strongest for scenarios whose major outputs are projections of one quantitative variable (or very few), and weakest for complex multivariate scenarios with substantial qualitative or narrative elements. The controversy over probabilities in SRES reflected in part different perceptions of what type of scenarios these were. SRES initially followed a storyline-based process and rejected quantification of probabilities on that basis. Subsequent efforts, however, consisted predominantly of developing quantitative emissions projections and neglected further development of the storylines. Moreover, with a few significant exceptions, subsequent applications of the scenarios have principally used their emissions figures, sometimes together with population and GDP, and made little or no use of the underdeveloped storylines that lay behind them. The controversy over quantitative probability in this case suggests that when quantitative projections are a major output of a scenario exercise, developers may have a responsibility to go further in characterizing the likelihood of the resultant emissions intervals than would be appropriate for the more complex underlying storylines.

Moreover, even for rich narrative scenarios, the arguments against rendering probability judgments are strongest when the exercise is produced for a small number of users with similar responsibilities and concerns. In such a setting, intensive interaction between scenario developers and users can provide whatever additional detail about, or confidence in, the scenarios that users may require to benefit from the scenarios. When scenarios serve potential users who are more numerous and diverse, however, such intensive interaction is not possible. As a result, the value of explicit likelihood judgments increases. To the extent that future global-change exercises continue to strengthen their qualitative aspects and the integration between qualitative and quantitative – valuable directions for future efforts – they should still seek to move further toward explicit characterization of likelihood than has been done thus far.

BOX 4.8. The Global Business Network Abrupt Climate Change Exercise

In 2002, the Office of Net Assessments (ONA), a small strategic planning office in the Office of the US Secretary of Defense, asked the Global Business Network (GBN), a consulting firm expert in scenario methods, to develop a scenario of potential national-security implications of abrupt climate change. This request was stimulated by widespread scientific interest at the time in abrupt climate change, particularly shifts in North Atlantic circulation, including a 2002 report by the National Academy of Sciences.[192] In addition, several scientific papers had reported changes in Atlantic circulation and salinity that some scientists thought might indicate impending larger disruption, as well as new evidence of rapid climate shifts in the past.[193]

GBN staff developed the scenario by reviewing scientific literature and informally consulting with climate and ocean scientists.[194] They reviewed three past climate events of diverse severity and decided to base their scenario on the one in the middle, the century-long period of strong cooling 8,200 years ago. Coming after an extended warm period, this event brought cooling of about 5 °F over Greenland, with cold and dry conditions extending around the North Atlantic basin and substantial drying in mid-continental regions of North America, Eurasia, and Africa.[195]

For their future abrupt-change scenario, the authors constructed a path of climate change to reach these conditions by 2020. The pathway involved rapid warming through 2010, as high as 4 – 5 °F per decade in some regions,[196] followed by a rapid turn to cooling around 2010, as melting in Greenland freshens the North Atlantic and substantially shuts down the thermohaline circulation. By 2020, hypothesized conditions have approached those of the 8,200-year event – cooling of 5 °F in Asia and North America and 6 °F in Europe, with widespread drying in major agricultural regions and intensification of winter storm winds. The authors acknowledge that the scenario pushes the boundaries of what is plausible, both in the rapidity of changes and in the simultaneous occurrence

continued on next page

---

[192] NRC 2002.

[193] See, e.g., Dickson et al. 2002, Hansen et al. 2001, Gagosian 2003, Curry and Mauritzen 2005, Fairbanks 1989.

[194] Global Business Network 2004.

[195] Alley et al. 1997.

[196] Note: these regional projections are 5-10 times faster than the IPCC's projected global 21st-century warming.

BOX 4.8, continued from previous page.

of extreme changes in multiple world regions. They contend that this is defensible and useful, however, for an exercise focused on sketching the nature of challenges posed by a plausible worst case.[197]

The socio-economic and security implications of the climate scenario were developed judgmentally, in consultation with ONA. Incremental changes are projected for the first 10 years, with general increase in environmental stresses and approximate maintenance of present disparities between rich and poor countries. After 2010, catastrophic cooling in Europe and drying of major agricultural regions worldwide brings widespread shortages of food, due to decreased agricultural production; of water, due to shifted precipitation patterns; and of energy, due to shipping disruptions from increased sea ice and storminess. These shortages produce 400 million migrants over the period 2010-2020, as desperate scarcity generates violent conflict in Europe, Asia, and the Americas. Extending their speculation on security implications into the 2020s, the authors hypothesize widespread southward migration of Europeans and near-collapse of the European Union, sustained conflict in East and Southeast Asia including struggles between China and Japan over access to Russian energy supplies, and increasing political integration of a fortress North America to manage security risks and refugee flows.

Controversy and criticism

The project was completed in October 2003, its report published in February 2004 and reported in Fortune Magazine the same month.[198] A few weeks later, the *London Observer* claimed to have obtained the report secretly and used the scenario to criticize US refusal to join the Kyoto Protocol.[199] Subsequent news coverage took up the theme that the report was secret or suppressed, and suggested the reason was that the scenario called for more urgent action on climate change.[200] In the resultant controversy, ONA stated — correctly — that the report did not represent US policy, but was merely a speculative consultant's study. Although the controversy subsided after a few weeks, interest and concern about the possibility of abrupt climate change, although not of this precise character, have continued to grow.[201]

This scenario is a sketch of an abrupt climate-change event, with little fine-scale detail about the hypothesized changes or underlying reasoning and no attempt to suggest how likely or unlikely such an event is. Rather, it seeks a preliminary answer to the question, what might the worst case look like? Such questions are more often posed in security studies than other fields, because of the unique nature of responsibilities of military organizations — responding to diverse, novel, unknown threats with extremely high cost of failure. Many climate-change decision-makers could likely benefit from such upper-bound scenarios too, but this exercise is the only example of a worst-case scenario produced for climate change. Major official assessments have focused overwhelmingly on average or best-guess projections.

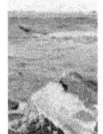

But the response to this report vividly illustrates the risks of worst-case or extreme scenarios. Produced in consultation with a sophisticated user — and in this case, one closely connected to senior decision-makers — who thoroughly understands the outer-bound nature of the underlying assumptions, they can be valuable devices for preliminary risk assessment and threat identification. But in a wider and polarized policy debate they are hard to explain and may be misunderstood or misrepresented. Attempting to manage the process through secrecy appears counterproductive, foregoing the potential value such analyses could provide to multiple decision-makers. More promising might be to integrate extreme-case scenarios explicitly into analyses that also present multiple mid-range scenarios.

---

[197] GBN 2004a:7, 2004b; Schwartz interview.

[198] Stipp 2004 (released, January 26, 2004).

[199] *London Observer* 2004.

[200] *San Francisco Chronicle* 2004, *Providence Journal* 2004.

[201] E.g., Alley 2004 cites it as a useful worst-case assessment.

Conclusions and Recommendations

This section presents our conclusions regarding the present state of development and use of scenarios for climate-change applications, and some recommendations for specific changes or initiatives to advance current practice to make scenarios more useful.

Before doing so, we briefly reprise some key definitional points, because uses of the term scenarios are so divergent. We have defined scenarios as descriptions of future conditions produced to inform decision-making under uncertainty. This definition distinguishes scenarios from assessments, models, decision analyses, and other decision-support activities. Scenarios may be developed and used in conjunction with these – for example, scenarios can provide descriptions of potential future conditions used as inputs to such activities – but are not identical to these, and not alternatives to them.

We have also distinguished scenarios from other types of future statements intended to inform decisions, such as projections, predictions, and forecasts. Relative to these, scenarios tend to be more multivariate (but still schematic), tend to be developed in groups, and tend to presume lower predictive confidence. The last condition is the case in part because scenarios tend to be used in situations where the basis for forecasting is less established because of deeper uncertainties, or for situations that pertain to further in the future beyond the range for which there is high confidence in specific projections, even contingent ones.

Having distinguished scenarios from these related activities, we consider a broad set of scenarios of diverse characteristics and uses, including simple and complex scenarios, quantitative and qualitative scenarios as well as various combinations of the two, and scenarios whose primary use and interpretation is positive or normative. Where we intend our conclusions and recommendations to apply to only certain types or uses of scenarios, we state this explicitly. Unless stated otherwise, they pertain to all types of global-change scenarios we are considering.

5.1 USE OF SCENARIOS IN CLIMATE-CHANGE DECISIONS

Scenarios can make valuable contributions to climate-change decision-making. Many of the decisions that will comprise society's response to climate change – whether mitigation, adaptation, or other responses – involve high stakes, deep uncertainties, and long time horizons. Scenarios can help inform these decisions by structuring present knowledge and uncertainty, prompting critical examination of present assumptions and practices, stimulating new insights, identifying key pitfalls and opportunities, or providing a framework for the assessment of particular decisions. For some decisions, which involve irreversible near-term commitment to choices whose consequences extend over a horizon involving substantial uncertainties, some form of scenario-based reasoning may be essential.

There is a big gap between the use of scenarios in current practice and their potential contributions. Despite their evident value and capability, many climate-related decisions that could benefit from scenarios (e.g., many decisions regarding long-term management and investments in climate-sensitive areas such as freshwater systems or coastal zones) are not using them. Indeed, many such decisions are still being made without considering climate change at all. Conversely, many climate-change scenarios have only weak and indirect connections to practical decisions related to climate-change mitigation or adaptation.

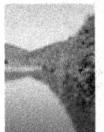

Interest in considering and using climate-change scenarios is sharply increasing. There is increasing interest in considering climate-change scenarios in diverse decision and planning processes. This trend is strongest for planning and decisions concerned with climate-change impacts and adaptation. The trend reflects advances in scientific understanding of climate change, gradual maturation of models and analytic tools, and increasing recognition by decision-makers of the potential importance of climate change. Given the high general concern about climate change and the advance

of background scientific knowledge, we expect this trend to continue, and to broaden to other types of climate-related decisions.

Scenarios of global emissions and resultant changes in atmospheric trace-gas concentrations and climate are a core requirement shared by many diverse climate-related decisions. Although climate-change decision-makers and their particular information needs are highly diverse, many will need scenarios of global emissions and resultant climate change, and many more will need information that depends on these. Consistent scenarios of global emissions and climate change, provided centrally at the national or international level, can serve these diverse needs – if they are presented with enough transparency and documentation of their underlying reasoning and assumptions.

Beyond global emissions and resultant climate change, decision-makers' needs from scenarios are highly diverse. Different climate-change decision-makers will have highly variable needs from scenarios, in the factors and variables included, the time and spatial scale at which they are provided, and the nature of uncertainties represented. The means for meeting these additional needs will likely be diverse, too. Some will call for separate, specialized scenario production capabilities. A major distinction in scenario-related needs can be drawn between impacts and adaptation managers, mitigation policy-makers, and energy resource and technology managers.

Impacts and adaptation managers need scenarios that project impacts relevant to their specific responsibilities, and the major determinants of vulnerability and adaptive capacity. Impacts and adaptation managers include both national officials and others responsible for more specific domains of impact. These decision-makers need climate-change scenarios, driven by specified global emissions scenarios, to provide information about potential climate-related stresses on their areas of responsibility. In addition, they need other environmental and socio-economic information specific to their areas of responsibility, at appropriate spatial and temporal scales. Meeting these needs will require both easy access to centrally produced climate scenario information with associated tools and support, and development of decentralized capabilities for developing and applying additional scenario-related information. Many of these specific information needs are likely to be similar in character for many particular locations and types of impact.

Meeting information needs for impacts and adaptation requires a cross-scale organizational structure. These decisions' combination of centralized and decentralized information needs suggests the need for a linked network of institutions at national and sub-national levels to develop scenarios. Such a structure would combine central provision of globally consistent emissions and climate change scenarios; decentralized elaboration of these scenarios with additional variables required for regional impacts and adaptation analyses; and provision of tools and resources to support development and use of scenarios.

Scenarios for impact and adaptation managers should be based on emissions assumptions that include a likely range of mitigation interventions, now and in the future. The emissions assumptions underlying scenarios for impacts managers should be based on the likely range of future global emissions trajectories, including explicit assumptions about what degrees of further mitigation effort are likely over time. This will typically imply a narrower range of emission futures than is considered in scenarios to support mitigation decisions.

Mitigation policy-makers need scenarios that project alternative emissions trends in their own jurisdiction and others, and the major factors that will influence mitigation opportunities, constraints, and costs. Mitigation policy-makers are usually officials who make national policy and participate in international negotiations, but this group also includes sub-national officials

when they share mitigation responsibilities or undertake mitigation initiatives. Serious pursuit of greenhouse-gas mitigation will require major policy innovations that carry significant risks of many kinds, including the effectiveness and cost of the policies but also their effects on government budgets, competitiveness of particular industries, opportunities for national technological capabilities, etc. Decision-makers considering such policies will need scenarios of global and national emissions trends, resultant climate change, and aggregate impacts. In addition, they will need to consider many factors specific to their jurisdictions – e.g., national policies, institutions, economic structure, technological capabilities, and the detailed structure of national emissions – and information about the policy environment for their choices, including alternative scenarios of other nations' mitigation strategies, international mitigation decisions, and implementation and compliance.

Scenarios for mitigation decisions should include a wide range of baseline emissions assumptions and should not prejudge the likely level of mitigation effort. Scenarios used to inform mitigation decisions should consider the full range of potential mitigation choices on the agenda, defined relative to baseline assumptions that, as much as possible, reflect only efforts already enacted or committed, including a range of reasonable assumptions about implementation and compliance. This assumption typically implies a wider range of emissions futures than is considered in scenarios to support impacts and adaptation decisions.

Mitigation decision-makers can use target-driven scenarios for backcasting. Mitigation decision-making may also benefit from scenarios that impose explicit future environmental targets such as limits on emissions or atmospheric concentrations, together with assumptions about policy and implementation elsewhere, and reason backwards to explore alternative paths to, and implications and requirements of attaining that goal, including feasibility, costs, and tradeoffs. These must be defined in ways relevant to the level of decision-making being informed, i.e., alternative national targets to inform national policy-making, in the broader context of alternative global baselines or global targets.

Informing mitigation decisions requires capacity for scenario development at the national level. While core scenarios of global emissions and climate-change can provide some of the required input into mitigation decisions, these decisions require additional information that must be provided at the national or sub-national level where the decisions are being considered, generated in consultation with relevant decision-makers or their surrogates.

Energy resource and technology managers need scenarios that represent the political and economic environment for energy investments, including mitigation policies. Energy resource and technology managers concerned with private responses to mitigation policy primarily need scenarios that represent alternative policy regimes. Emissions and climate change underlie these as influences on policy decisions, but do not capture the most important uncertainties for these decision-makers. While many actors may wish to generate these scenarios privately to keep their assumptions and analyses confidential, there may also be value in multi-party collaborative scenario-building exercises in which today's policy-makers and corporate planners jointly examine what range of policy, economic, and energy regimes is plausible or likely over the 30- to 50-year time horizons relevant for investment and technology-development decisions.

Scenarios must be periodically revised and updated. For all types of decisions and decision-makers, developing scenarios, applying them to inform decisions, and refining scenario methods, are iterative processes. Limitations to present scenarios or methods do not in general justify delaying consideration of such decisions, any more than scientific uncertainties do. Still, scenarios must be periodically updated,

based on new knowledge, experience, and priorities, as well as further developments in scenario-related methods. Such updates are needed much more frequently than the decision time horizons.

5.2 USE OF SCENARIOS IN CLIMATE-CHANGE ASSESSMENTS

Large-scale, official assessments are the major use for scenarios at present and are likely to remain an important use. Large-scale, official assessments represent the most prominent demand for climate-related scenarios at present, and are likely to remain major users, particularly for coordinated scenarios of global emissions and resultant climate-change.

Within assessments, scenarios mainly serve to support further analysis, modeling, and assessment. When scenarios are used in assessments, certain users are clearly identified: e.g., climate modelers are major users of emissions scenarios, while impacts assessors and modelers are major users of climate-change scenarios. These users have specific scenario needs, and close consultation is possible between scenario producers and users to meet these needs. Substantial progress has been made in providing useful scenarios for these groups, at both the national and international level. These efforts should be continued and expanded.

The presentation of scenarios in assessments leads to many additional, unforeseen uses. Scenarios presented in large-scale assessments gain prominent dissemination that results in their being put to many uses their developers did not foresee. Scenarios should pursue clarity of documentation and transparency about underlying reasoning and assumptions, to improve the ease of use and reduce the risk of misunderstanding in such derivative uses, although they cannot anticipate all information needs of an open-ended set of diverse potential uses.

In assessments, scenarios can strongly influence issue framing. Also because of

their prominent dissemination, scenarios presented in major assessments can exercise substantial influence over the framing of policy discussions or provide simple, widely used metrics of the seriousness of the issue. They may consequently exercise broad influence over many decisions that depend upon such aggregate perceptions of seriousness. The prospect of such influence further heightens the responsibility for transparency in production of scenarios.

Scenarios contain unavoidable elements of judgment in both their production and use. Although they draw on relevant data, knowledge, and analysis, scenarios inevitably contain elements of judgment. In addition to putting serious responsibilities onto scenario developers, this implies that there is no authoritative way to resolve arguments over whether a scenario is plausible or not. When a wide enough range of potential futures is considered, some scenarios are likely to draw criticism, in part motivated by opposition to their foreseeable implications for action. Any scenario can be attacked as unreasonable, speculative or unlikely, and close enough scrutiny of any scenario can usually reveal inconsistencies, but these do not provide sufficient basis for excluding a scenario from consideration. Indeed, scenarios designed to represent extreme events, or to lie near one end of a distribution of potential outcomes, should by definition appear unlikely. The most productive response to such criticisms lies in transparency about the process, reasoning, and assumptions used to produce scenarios. Such transparency can shift arguments to underlying uncertainties, and help limit biases in the production of scenarios.

5.3 CHARACTERISTICS OF "CORE" EMISSIONS AND CLIMATE SCENARIOS

Centrally provided scenarios of emissions and climate change can help inform mitigation and adaptation decisions at national and sub-national scale, but these will usually require additional information as well. Central scenarios can provide information about trends in world emissions, underlying socio-economic conditions at the scale of major world regions, and the large-scale pattern of global policy response. They can also provide access to climate-model scenario output, plus tools, data, and support for producing finer-scale scenario information needed for particular impact and adaptation applications. Mitigation and adaptation decisions and associated assessments at national or smaller spatial scale will need more detailed and finer-scale climate and socio-economic information than can be provided by centralized scenarios, so these must be extended and/or modified by national and sub-national scenario processes.

Scenarios of emissions and resultant climate change should be global in scope and century-scale in time horizon. Core emissions and climate-change scenarios should specify major climate-relevant emissions and other perturbations, globally and for major world regions. They should extend over a time horizon of at least 100 years (including some that extend 200-300 years to support assessments of sea level rise), with interim results at roughly decadal resolution.

Emissions scenarios of several distinct logical types will be needed to serve diverse purposes. These will include some combination of alternative baselines, alternative levels of incremental stringency of mitigation effort, and specified future targets to support backcasting and feasibility analysis.

For some uses, emissions scenarios should be coupled to explicit scenarios of alternative socio-economic futures. For these scenarios, the range of potential socio-economic and policy futures considered should be wider than has been considered to date, including scenarios of policy failure and conflict, and a wide range of stringency and timing of mitigation effort. For example, what if development stagnates in major world regions? What if world emissions grow sharply for several decades with little control effort, followed by a subsequent shift to stringent mitigation efforts? What if part of the world makes a lot of effort and part makes very little? Considering such varied future histories is crucial for considering long-term risks and opportunities from major mitigation choices.

Scenarios should reflect various explicit degrees of coordination, depending on their intended uses. Some uses will require groups of simple coordinated scenarios to provide standardized inputs for downstream modeling and analysis – e.g., standard emissions scenarios as inputs to climate models and standard climate scenarios as inputs to impact assessments – for exploring present uncertainties and tracking developments of knowledge over time. Other scenarios should be based on multiple models using common input assumptions. Non-standardized scenarios produced at the initiative of researchers and modelers should also be produced, which explore alternative assumptions or meet specific user needs, provided these meet basic standards of quality control, transparency, and documentation.

Some scenarios should seek to link qualitative and quantitative elements. Some scenarios of socio-economic conditions, whether produced to support global emissions scenarios or impacts assessments, should include both qualitative and quantitative elements and sustained analytic efforts to link the two. Qualitative or narrative scenario elements can provide a vehicle to explore major historical uncertainties with large implications for global emissions, climate change, and vulnerability to climate impacts; provide a coherent rationale and logical structure to connect assumed trajectories for multiple variables, including both quantitative and qualitative ones; and provide guidance to other analysts or users who may wish to extend the scenarios by elaborating additional detail. Achieving these benefits will require more sustained effort to integrate model-based projections of quantitative variables with qualitative and narrative scenario elements, to iterate between these, and to critically examine each element in light of the other, than has

been made thus far. These efforts should seek to connect alternative qualitative and narrative scenarios not just to alternative parameter values in quantitative models, but also to alternative forms of causal relations and model structures. Generating multiple alternative model quantifications based on the same narrative and associated causal logic is one promising route to enriching understanding of uncertainties in key quantitative variables such as future economic output and emissions.

5.4 SCENARIO PROCESS: DEVELOPER-USER INTERACTIONS

There is value in collaboration between scenario developers and users, particularly at the beginning and ending stages of a scenario exercise. The appropriate degree and means of this collaboration vary substantially among scenario exercises. User engagement is most important in the initial scoping and design of a scenario exercise, and in the evaluation and application of the scenarios generated. The value of user engagement in details of scenario development, quantification, elaboration, and checking, depends on the specific case.

The ease of achieving such collaboration and its value are likely to be greater when scenario users are clearly identified, few in number, and similar in their interests and perspectives. When potential scenario users are identified, relatively few, and relatively homogenous, close and intensive collaboration between users and developers is likely to be most productive. When potential users are numerous and diverse, intensive engagement may be infeasible and more structured processes for consultation, representation, and information exchange are needed. While progress has been made in new methods to allow larger numbers to participate in scenario exercises, further development of such methods is needed.

5.5 COMMUNICATION OF SCENARIOS

Effective communication of scenarios is essential, in forms useful to audiences of diverse interests and technical skills. Scenarios must be communicated effectively to their potential users, including both technical and non-technical audiences. In addition to the contents or outputs of scenarios, communication should include associated documentation, tools, and support for their use. Various methods should be used to promote broad dissemination of scenario information; for instance, presentations, reports, websites, and centralized data distribution centers. To facilitate user understanding of results, various methods should be used to communicate numerical and technical information, including multiple tabular, summary, and graphical formats, ideally with user-interactive capabilities.

Transparency of underlying reasoning and assumptions is crucial. Scenario communication should include transparent disclosure of underlying assumptions, models, and reasoning used to produce the scenarios, to support the credibility of scenarios, to alert potential users to conditions under which they might wish to use or modify them, and to inform criticism and improvement of scenarios. This should include explicit identification of the major uncertainties represented in each scenario and the sources of underlying information, whether drawn from the scientific literature, formal expert-elicitation exercises, or informal judgments of the scenario team. It is possible in virtually all cases to formulate simple, accessible, honest descriptions of why a scenario was undertaken, why it was necessary, what was done, how and why, and why it merits respect as a reasonable judgment.

5.6 CONSISTENCY AND INTEGRATION IN SCENARIOS

Any scenario should be internally consistent in its assumptions and reasoning, to the extent this can be established given present knowledge. Carefully pursuing consistency within individual scenarios can be an intensive and time-consuming process, but is crucial to avoid problems that can discredit a scenario exercise.

In scenario exercises that use multiple models to explore potential uncertainties in future conditions, consistency between models should be pursued primarily through coordination of inputs, not outputs. Use of multiple models in parallel to produce alternative descriptions of future conditions can improve understanding of uncertainties, if models are run under consistent assumptions about exogenous inputs. Forcing models to generate consistent trajectories for endogenous outputs poses several risks, including suppressing variation from alternative causal structures that could provide valuable insights into uncertainties, and encouraging over-confidence from spurious precision. For quantities that are exogenous in some models and endogenous in others, the appropriate treatment varies case by case, but it is not generally desirable to force multiple models to convergent values of such variables without more detailed examination of the underlying uncertainties.

Imposing consistent outputs in multi-model exercises can be useful, however, when these outputs represent common goals for policy evaluation. For example, consistent constraints on some environmentally relevant target such as emissions, atmospheric concentrations, or radiative forcings, can be used to examine inter-model uncertainties in the technological, economic, and resource conditions associated with meeting the specified targets.

Transparency in reporting scenario and model differences as well as underlying assumptions and reasoning can help mitigate the effects of inconsistencies among scenarios. Ideally, multiple scenarios in an exercise should differ only on those elements intentionally chosen to distinguish them, and be consistent in all other factors. However, this is not always possible, particularly when scenarios are produced using different models. Pursuing maximal transparency about the models, assumptions, and reasoning underlying each scenario – perhaps by issuing detailed diagnostic reports that include explicit discussion of points of weakness, uncertainty, and disagreements, and the means used to resolve them – can mitigate any resultant confusion.

5.7 TREATMENT OF UNCERTAINTY IN SCENARIOS

Some scenario exercises should include more explicit characterization of likelihood judgments than has been practiced so far. The advantages of being more explicit about the probability judgments that underlie scenario exercises are likely to outweigh the disadvantages. Such specification should be pursued further than has been done in major global-change scenario exercises to date, although not necessarily in all scenario exercises. The means available to express these judgments are of widely varying specificity, ranging from agreed terminology[202] to explicitly quantified probability distributions. All such judgments should include explicit acknowledgement of their inevitably subjective elements and appropriate caveats to help users avoid mistaking them as objectively true.

Explicit probability judgments are easiest to produce and least controversial in scenarios generated using quantitative models of climate change or specific impact domains. Scenarios generated using such models can be conditioned on specific assumed values for socio-economic inputs such as emissions, and can represent explicitly and quantitatively the effects of specified variation in initial conditions or unknown parameter values. These devices are also available, although in less widespread use, in economic models used to project emissions. These devices aid in constructing distributions of key quantitative characteristics, such as measures of

global or regional climate, or of prominent quantitative impact measures, such as changes in river flows or sea level, although they neither capture all relevant uncertainty nor avoid the inevitably subjective nature of such probability judgments. Explicit probability judgments are more difficult and controversial when they involve socio-economic factors for which quantitative models are not available, and that do not depend in well understood ways on identified quantitative parameters. Such factors include major technological innovations, large-scale changes in attitudes or norms, or policy response.

Attempting to include explicit probability judgments is likely to be most useful and successful when key variables are few, quantitative outcomes are needed, and potential users are numerous and diverse. The case for assigning explicit likelihood judgments is strongest when scenarios' most salient components are quantitative projections of a few key variables, such as emissions or average temperature change over the globe or some region, because the technical barriers to assigning probabilities are least severe in this case. The case is strongest when a primary purpose of the scenario exercise is to provide inputs to other quantitative assessment activities, or to inform decisions that primarily depend on one or a few key quantitative variables, because such uses are most likely to require probability judgments. The case is strongest when the set of potential scenario users and uses is large and heterogeneous, because this situation provides the least opportunity for informal communication of implicit judgments of likelihood or priority through intense, sustained collaboration between scenario developers and users.

Attempting to include explicit probability judgments is likely to be least useful and successful when scenarios specify multiple characteristics, including prominent narrative or qualitative components; when the purpose of a scenario exercise is sensitivity analysis or heuristic exploration;

---

[202] See, e.g., the consistent uncertainty language proposed for IPCC reports by Moss and Schneider 2000.

and when potential users are few, similar, and known. When scenarios are primarily construed as rich, qualitative narratives that present major alternative historical and socio-economic trajectories, the technical obstacles to explicit probability assignment are greatest and the likely confidence in scenario developers' subjective probability judgments lowest. When the main purpose of a scenario exercise is to stimulate critical or creative thought, to probe the limits of a model or decision strategy through sensitivity analysis, or to explore ways of meeting a specified target, explicit probability assignment provides little or no benefit. When users are few, similar, and specifically identified, they can be intensively involved in scenario production, allowing effective informal communication of likelihood judgments without stating them explicitly. Under these conditions, scenario exercises can also be structured to engage users in the potentially instructive activity of assigning and discussing their own probability judgments, rather than putting that responsibility exclusively on the researchers or analysts developing scenarios.

Centrally provided scenarios of global emissions and climate change should attempt to include explicit probability judgments. Because of the large, diverse set of users for these scenarios, explicit likelihood judgments should be provided for ranges of key quantitative outputs, including global emissions and global-average temperature change. Scenarios should typically include several paths that span a wide range of judged uncertainty, e.g., 95 percent to 99 percent. The associated probability judgments may include some that are unconditional and some that are conditioned on specific assumed prior conditions. Such estimates should be provided by multiple groups using diverse methods. As for all such probability judgments, their unavoidably subjective nature and the specific assumptions on which they are conditioned should be stated explicitly and prominently

Providing explicit probability judgments allows scenario users to choose whether to use them or not. Some users may choose to use these explicitly in their subsequent analysis or decision support, while others may use them only to help decide which scenarios to use, and still others may disregard them entirely. Users may select a different group of scenarios or a different subset of the uncertainty range for various reasons, including differences in risk aversion, differences in the scope of their decision authority, or differences in their assumptions about decisions by other actors (present or future). Presenting complete descriptions of scenarios together with underlying assumptions and reasoning, including probability judgments, preserves all these options for users.

Scenario exercises should give more attention to extreme cases. Some uses of scenarios require consideration of low-probability, high-consequence extreme cases, such as loss of a major continental ice sheet or major changes in meridional ocean circulation. Consequently, such scenarios should be included in large, general-purpose scenario exercises producing emissions or climate-change scenarios, together with more likely middle-case scenarios. When extreme scenarios are included in an exercise, it is especially critical to be explicit and transparent about the reasoning and assumptions underlying each scenario, and scenario developers' judgments of relative likelihoods.

In addition to enhancing the utility of scenario outputs, probabilistic methods can contribute throughout the scenario development process. Developing scenarios requires making many judgments about unknown characteristics and developing many arguments and pathways to link these. Scenarios based on quantitative models typically require specifying many exogenous inputs and parameters. Even narrative scenarios require specifying values of multiple characteristics, both qualitative and quantitative. Explicit discussion of uncertainties and associated probabilities can help structure

and facilitate many aspects of the scenario development process, including deciding appropriate ranges of variables to consider, defining boundaries of what outcomes are considered plausible, elaborating associated causal mechanisms and linkages, discussing and integrating knowledge and judgments from multiple participants, and clarifying disagreements. Explicit conversation about probabilities can support insights throughout these processes, in addition to supporting communication of scenario judgments to users.

5.8 EXPANDING AND SUSTAINING CAPACITY FOR PRODUCTION AND USE OF SCENARIOS[203]

Present scenario capacity is inadequate. Although scenario-based reasoning is required for many aspects of global change assessment and decision support, the present capacity to produce, disseminate, apply, evaluate, and adapt scenarios is inadequate. There has not been enough continuity to enable effective learning, because scenarios are typically produced *de novo* for each major application. There has not been enough transparency about methods, reasoning, and assumptions. Constructing integrated scenarios and exploring alternative methods has been difficult, in part because scenario exercises have tended to be dominated by use of quantitative models, separated along disciplinary lines. Inadequate resources have been devoted to methods development, for scenarios and related decision-support tools. Finally, there has been no systematic evaluation and critique of scenarios or their application.

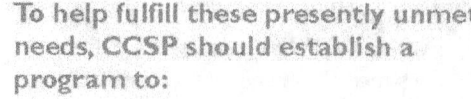

To help fulfill these presently unmet needs, CCSP should establish a program to:

• *Commission scenarios for use in assessments and decision-support activities.* This task includes facilitating agreement among relevant producers and users on standard scenarios in cases where multiple assessment activities require standard scenarios for comparability, and convening and supporting a diverse collection of more extensive and detailed scenario-related exercises, by multiple groups using a wide range of models and approaches.

• *Disseminate scenarios with associated documentation, tools, and guidance materials.* This task includes multiple forms of support and program-building for diverse groups seeking to apply, modify, and extend existing scenario-based information at various regional and sectoral scales, through providing data, models, tools, resources, and associated documentation and technical support, in multiple forms and through multiple media.

• *Commission various groups to evaluate scenarios and their applications and to develop improved methods.* This task includes defining and promulgating standards for quality control – which, given the need for diverse approaches, would principally concern matters of process such as transparency, documentation, evaluation, and dissemination of results and supporting information. The task also includes broad efforts to develop scenario-related skills, tools, and methods, e.g., by providing resources for methods development and evaluation projects; conducting and establishing procedures for evaluation of particular activities and programs; and convening workshops,

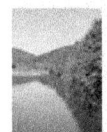

---

[203] Recommendations made in this report regarding programmatic and organizational changes, and the adequacy of current budgets, reflect the judgment of the report's authors and the CPDAC and are not necessarily the views of the U.S. Government.

conferences, etc., to evaluate progress overall, or in particular methodological areas.

• *Archive results and documentation related to all these tasks, to provide historical perspective and institutional memory for future scenario-related activities.* This task includes preserving for retrospective scrutiny a wide range of materials: not just the methods, contents, and results of scenario exercises, but also the progressive evaluations of particular activities and approaches, and of the entire program. In its ongoing development and evaluation of methods, the program should not draw rigid boundaries between scenario development and application and other related methods of assessment and decision support for long-term global change issues.

Several conditions in the design and management of this new program would be required to ensure its effectiveness.

• *The program should build and maintain strong connections with outside relevant expertise, and analytic and modeling capability.* While the program must develop enough internal expertise to be a full participant in debates over scenarios and assessment methods, it cannot and should not attempt to impose a unilateral vision of preferred scenarios, methods, or approaches. Rather, it must build and maintain close collegial connections with outside networks of researchers and analysts in multiple fields of expertise, including emissions modelers, climate scientists and modelers, impacts researchers, and resource managers. These relationships would be facilitated by establishing governing mechanisms, such as a senior advisory board, drawn from the broad communities of researchers, modelers, and ana-

lysts who are developing and using scenarios and related methods. Although established as a US national program, it must also support, collaborate, and coordinate with parallel activities in other nations and internationally, and with relevant sub-national activities.

• *The program should integrate and balance goals and criteria related to scientific and technical quality, and those related to utility and relevance to users.* This balance is needed for the program to support promising but speculative activities, to encourage creativity and diversity of approach, to avoid being captured by any particular discipline or modeling approach, and to be willing to make and explain judgments about quality and promise that reflect both scientific and practical considerations. To achieve this, the program needs broad discretion over the type of projects supported, including sponsoring fairly sharply targeted activities, supporting speculative activities, and investing to develop and assess capabilities that do not yet exist.

• *The program should be insulated from political control.* For the scenarios and analyses based on them to be perceived as credible by their diverse users, the program needs enough insulation from political control, at both the national or international level, to prevent scenarios from becoming proxies for conflict over near-term policies, and to allow exploration of the implications of alternative futures that represent plausible risks but that some political actors would find objectionable.

• *The program should strive for maximum transparency in its own activities, in addition to demanding it from activities it supports.* The program should strive for maximal transparency regarding inputs, mod-

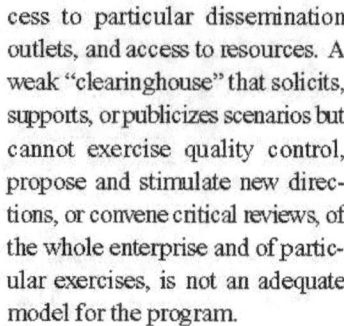

els, assumptions, and reasoning employed in developing scenarios, as well as any significant disagreements that arose and how they were resolved and any remaining weaknesses recognized by the developers. The broader and more diverse the collection of intended uses and users, the more crucial is transparency of the scenario-production process – because different users may require scenarios produced using different underlying assumptions, and they must be able to track the underlying logic to exercise this choice. This would enhance credibility in the scenario-development process. While calls for such transparency are widely made, experience suggests it is difficult to achieve, particularly for such matters as disagreements or recognized weaknesses that may risk professional embarrassment. Still, achieving more transparency and more widely informed debate on such matters is essential for advancing scenario methods.

• *The program will require the authority and resources necessary to articulate and promulgate standards for transparency, consistency (e.g., of units and formats), and quality control.* This task involves facilitating discussions among the community and formulating persuasive guidelines and supporting arguments. It also requires use of incentives such as seals of approval, access to participation in particular processes, ac-

cess to particular dissemination outlets, and access to resources. A weak "clearinghouse" that solicits, supports, or publicizes scenarios but cannot exercise quality control, propose and stimulate new directions, or convene critical reviews, of the whole enterprise and of particular exercises, is not an adequate model for the program.

• *The program will require an adequate sustained resources.* The program must build and maintain a sophisticated analytic capability, and develop skills and institutional memory regarding prior experiences, successes, and failures. This requirement precludes the program being a series of *ad hoc* one-time activities or a part-time, unfunded burden imposed on people and organizations with other full-time responsibilities.

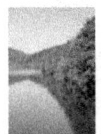

A1, A2, B1, B2, A1FI

    names of specific emissions scenarios published in the IPCC Special Report on Emissions Scenarios

CCSP   Climate Change Science Program

CCTP   Climate Change Technology Program

CFCs   chlorofluorocarbons

DDC   Data Distribution Center (IPCC)

DOD   Department of Defense (U.S.)

DWR   Department of Water Resources (California)

EMF   Energy Modeling Forum

ENSO   El Ni o/Southern Oscillation

EPA   Environmental Protection Agency

GBN   Global Business Network

GCM   general circulation model

GDP   gross domestic product

GFDL   Geophysical Fluid Dynamics Laboratory, US National Oceanographic and Atmos pheric Administration (NOAA), and climate model produced by this laboratory

GtC   gigatonnes (billion metric tons) of carbon

HadCM2   UK Hadley Centre climate model, Version 2

IIASA   International Institute for Applied Systems Analysis, Laxenburg, Austria

IPCC   Inter-governmental Panel on Climate Change

IS92   series of emissions scenarios produced by the IPCC in 1992, including specific scenarios named IS92a through IS92f

LRTAP   Convention on Long-Range Transboundary Air Pollution

MEA   Millennium Ecosystem Assessment

MEC   Metropolitan East Coast assessment of US National Assessment

MER   market exchange rates

NAPAP   National Acid Precipitation Assessment Program

NAST   National Assessment Synthesis Team

NRC   National Research Council (U.S.)

NYCDEP   New York City Department of Environmental Protection

NYCHP   New York Climate and Health Project

OECD   Organization for Economic Co-operation and Development

ONA   Office of Net Assessments (Office of US Secretary of Defense)

OTA   Office of Technology Assessment, United States Congress

PDO   Pacific Decadal Oscillation

PPP   purchasing-power parity

SRES   Special Report on Emissions Scenarios (IPCC)

TGICA   Task Group on Data and Scenario Support for Impact and Climate Analysis (IPCC)

UKCIP   United Kingdom Climate Impacts Program

VEMAP   Vegetation Ecosystem Mapping and Analysis Project

VOC   olatile organic compounds

REFERENCES

Alcamo, J., A. Bouwman, J. Edmonds, A. Gruebler, T. Morita, and A. Sugandhy, 1995: An Evaluation of the IPCC IS92 Emissions Scenarios." In: *Climate Change 1994: Radiative Forcing of Climate Change and an Evaluation of the IPCC IS92 Emission Scenarios* [Houghton, J.T., L.G. Meira Filho, J. Bruce, H. Lee, B.A. Callander, E. Haites, N. Harris, and K. Maskell (eds.)], Cambridge University Press, Cambridge, UK.

Allen, M., S. Raper, et al., 2001: Climate Change - Uncertainty in the IPCC's Third Assessment Report. *Science* 293, 5529, 430.

Alley, R.B., T. Sowers, P.A. Mayewski, M. Stuiver, K.C. Taylor, and P.U. Clark, 1997: Holocene climate instability: a prominent, widespread event 8,200 years ago. *Geology* 26, 6.

Alley, R.B., 2004: Abrupt climate change. *Scientific American*, November 2004, 62-69.

Arnell, N.W., M.J.L. Livermore, S. Kovats, P.E. Levy, R. Nicholls, M.L. Parry, and S.R. Gaffin, 2004: Climate and socio-economic scenarios for global-scale climate change impact assessments: characterizing the SRES storylines. *Global Environmental Change* 14, 3-20.

Barras, J., S. Beville, D. Britsch, S. Hartley, S. Hawes, J. Johnston, P. Kemp, Q. Kinler, A. Martucci, J. Porthouse, D. Reed, K. Roy, S. Sapkota, and J. Suhayda, 2003: *Historical and Projected Coastal Louisiana Land Changes: 1978-2050*. Open File Report 03-334, US Geological Survey, Washington, DC, USA.

Barney, G.O., 1981: *Global 2000 Report to the President*. Council on Environmental Quality, Washington, DC, USA.

Barrow, E., P. Gachon, and D. Caya, 2004: Creating climate change scenarios for Canada. In: *Climate Change and Variability in Canada: Past, Present and Future*. Environment Canada, Ottawa.

Berkhout, F. and J. Hertin, 2000: Socio-economic futures for climate impact assessment. *Global Environmental Change* 10, 165-168.

Berkhout, F. and J. Hertin, 2001: Foresight Futures 2001: Revised Scenario and User Guidance. Final Report of the Review and Revise Environmental Futures study, SPRU, Brighton, UK.

Berkhout, F., J. Hertin, I. Lorenzoni, A. Jordan, K. Turner, T. O'Riordan, D. Cobb, L. Ledoux, R. Tinch, J. Palutikof, M. Hulme, and J. Skea, 2001: Presentation of the UKCIP socio-economic scenarios for climate impact assessment. In: UKCIP, *Socio-Economic Scenarios for Climate Impact Assessment: A Guide to Their Use in the UK*, www.ukcip.org.uk/resources/publications/pub_dets.asp?ID=34.

Berkhout, F., J. Hertin, and A. Jordan, 2002: Socio-economic futures in climate impact assessment: using scenarios as learning machines. *Global Environmental Change* 12, 83-95.

Boer, G.J., G.M. Flato, M.C. Reader, and D. Ramsden, 1999a: A transient climate change simulation with historical and projected greenhouse gas and aerosol forcing: experimental design and comparison with the instrumental record for the 20th century. *Climate Dynamics* 16, 405-426.

Boer, G.J., G.M. Flato, and D. Ramsden, 1999b: A transient climate change simulation with historical and projected greenhouse gas and aerosol forcing: projected climate for the 21st century. Climate Dynamics 16, 427-450.

Bracken, P., 1977: Unintended consequences of strategic gaming. *Simulation and Games* 8, 283-318 (3 September).

Bracken, P., 1990: Gaming in hierarchical defense organizations. In: *Avoiding the Brink: Theory and Practice in Crisis Management* [Goldberg, A.C., D. Van Opstal, and J.H. Barkley (eds.)]. Brassey's, London.

Bretherton, F.P., K. Bryan, J.D. Woods, et al., 1990: Time-dependent greenhouse-gas-induced climate change. In: *Climate Change: The IPCC Scientific Assessment* [Houghton, J.T., G.J. Jenkins, and J.J. Ephraums (eds.)]. Cambridge University Press, Cambridge, UK.

Brewer, G.D., 1986: Methods for synthesis: policy exercises. In: Sustainable Development of the Biosphere [Clark, W.C. and R.E. Munn (eds.)]. Cambridge University Press, Cambridge, UK.

Brewer, G.D., 1990: Discovery is not prediction. In: Avoiding the Brink: Theory and Practice in Crisis Management [Goldberg, A.C., D. Van Opstal, and J.H. Barkley (eds.)]. Brassey's, London.

Brewer, G.D. and M. Shubik, 1979: The War Game: a Critique of Military Problem-Solving. Harvard University Press, Cambridge, MA.

Burkett, V.R., D.A. Wilcox, R. Stottlemeyer, W. Barrow, D. Fagre, J. Baron, J. Price, J. Nielsen, C.D. Allen, D.L. Peterson, G. Ruggerone, and T. Doyle, 2005: Nonlinear dynamics in ecosystem response to climate change: case studies and policy implications. *Ecological Complexity* **2**, 357-394.

California DWR, (Dept of Water Resources), 2005: *California Water Plan Update 2005*. Bulletin 160-05. Available at http://www.waterplan.water.ca.gov.

Castles, I. and D. Henderson, 2003a: The IPCC emissions scenarios: an economic-statistical critique. *Energy and Environment* **14(2-3)**, 159-185.

Castles, I. and D. Henderson, 2003b: Economics, emissions scenarios, and the work of the IPCC. *Energy and Environment* **14(4)**, 415-435.

CCSP, 2007: *Scenarios of Greenhouse Gas Emissions and Atmospheric Concentrations*. A report by the U.S. Climate Change Science Program and the subcommittee on Global Change Science Research [Clarke, L., J. Edmonds, J. Jacoby, H. Pitcher, J. Reilly, and R. Richels (Eds.)]. Department of Energy, Office of Biological & Environmental Research, Washington, DC, USA, 102 pp.

Chess, C. and K. Purcell, 1999: Public P and the environment: do we know what works? *Environmental Science & Technology* **33**, 2685-2691.

Cowling, E.B., 1992: The performance and legacy of NAPAP. *Ecological Applications* **2(2)**, 111-116.

Cook, C.W., 1939: *Scenery of Florida Interpreted by a Geologist*. The State Geological Survey, Tallahassee, FL, USA.

Curry, R. and C. Mauritzen, 2005: Dilution of the northern North Atlantic Ocean in recent decades. *Science* **308**, 5729, 1772-1774.

Cubasch, U., G.A. Meehl, G.J. Boer, R.J. Stouffer, M. Dix, A. Noda, C.A. Senior, S.C.B. Raper, and K.S. Yap, 2001: Projections of future climate change. In: *Climate Change 2001: The Scientific Basis* [Houghton, J.T., G.J. Jenkins and J.J. Ephraums (eds.)]. Cambridge University Press, Cambridge, UK.

Davis, G., 2003: Meeting future energy needs: choices and possibilities. *The Bridge* **33(2)**, 16-21.

de la Chesnaye, F. and J. Weyant (eds.), 2006: Multi-greenhouse gas mitigation and climate policy. *The Energy Journal*, special issue.

DeWeerd, H., 1967: *Political Military Scenarios*, P-3535. Rand Corporation, Santa Monica, CA.

DeWeerd, H., 1975. A contextual approach to scenario construction. *Simulation and Games* **5**, 403-414.

Dickson, B., I. Yashayaev, J. Meincke, B. Turrell, S. Dye, and J. Hoffort. 2002. Rapid freshening of the deep North Atlantic Ocean over the past four decades. *Nature* **416**, 832.

Doyle, T.W., R.H. Day, and J.M. Biagas, 2003: Predicting coastal retreat in the Florida Big Bend Region of the Gulf Coast under climate change induced sea-level rise. In: *Integrated Assessment of the Climate Change Impacts on the Gulf Coast Region - Foundation Document* [Ning, A.H., R.E. Turner, T. Doyle, and K. Abdollahi (eds.)]. LSU Press, Baton Rouge, USA.

Duncan, N. and P. Wack, 1994: Scenarios designed to improve decision making. *Planning Review* **22(4)**, 18-25.

The Economist, 2003a: Hot potato. The Intergovernmental Panel on Climate Change had better check its calculations. February 13.

The Economist, 2003a: Hot potato revisited: a lack-of-progress report on the Intergovernmental Panel on Climate Change. November 6.

EMF (Energy Modeling Forum), 2006: EMF 22: *Greenhouse Gas Stabilization*. Prospectus. www.stanford.edu/group/EMF/projects/EMF22/emf22prospectus.pdf.

Envision Sustainability Tools, 1999: Lower Fraser Basin QUEST Model Structure, Ver 1.0. Vancouver BC. www.sdri.ubc.ca/downloads/Models.pdf

Fairbanks, R.G., 1989: A 17,000-year glacio-eustatic sea level record: influence of glacial melting rates on the Younger Dryas and deep ocean circulation. *Nature* **342**, 7, 637-642.

Gagosian, R.B., 2003: Abrupt climate change: should we be worried? Presented to World Economic Forum, Davos, Switzerland, January 27.

GBN (Global Business Network), 2004a: *Executive Summary, An Abrupt Climate Change Scenario and Its Implications for US National Security*, Oct. 2003. www.ems.org/climate/pentagon_climatechange.pdf,

GBN (Global Business Network), 2004b: Abrupt Climate Change. Press Release, February. www.gbn.com/ArticleDisplayServlet.srv?aid=26231.

Giorgi, F., B. Hewitson, J. Christensen, M. Hulme, H. von Storch, P. Whetton, R. Jones, L. Mearns, and C. Fu, 2001: Regional climate information: evaluation and projections. In: *Climate Change 2001: The Scientific Basis* [Houghton, J.T., G.J. Jenkins, and J.J. Ephraums (eds.)]. Cambridge University Press, Cambridge, UK.

Godet, M. and F. Roubelat. 1996. Creating the future: the use and misuse of scenarios. *Long Range Planning* **29**, 2, 164-171.

Gosselink, J.G., 1984: *The Ecology of Delta Marshes of Coastal Louisiana: a Community Profile*. FWS/OBS-84/09, U.S. Fish and Wildlife Service, Washington, DC, USA.

Gough, C., N. Castells, and S. Funtowicz, 1998: Integrated assessment: an emerging methodology for complex issues. *Environmental Modeling and Assessment* **3**, 19-29.

Greenberger, M., G.D. Brewer, W.W. Hogan, and M. Russell. 1983: *Caught Unawares: The Energy Decade in Retrospect*. Ballinger, Cambridge, MA, USA.

Gregory, R. and T. McDaniels, 2005: Improving environmental decision processes. Appendix B in: NRC 2005, *Decision Making for the Environment*. National Academy Press, Washington DC, USA.

Grübler, A. and N. Nakicenovic, 2001: Identifying dangers in an uncertain climate. *Nature* **412**, 5-15.

Grübler, A., N. Naki enovi , J. Alcamo, G. Davis, J. Fenhann, B. Hare, S. Mori, B. Pepper, H. Pitcher, K. Riahi, H. Rogner, E.L. La Rovere, A. Sankovski, M. Schlesinger, R.P. Shukla, R. Swart, N. Victor, and T.Y. Jung, 2004: Emission scenarios: a final response. *Energy & Environment* **15**, 1, 11-24.

Hansen, B., W. Turrell, and S. Østerhus, 2001: Decreasing overflow from the Nordic Seas into the Atlantic Ocean through the Faroe Bank Channel since 1950. *Nature* **411**, 927-930.

Hausrath, A.H., 1971: *Venture Simulation in War, Business, and Politics*. McGraw-Hill, New York, USA.

Herrick, C., 2002: Atmospheric Science and the Constitution of Public Policy: The Case of the National Acid Precipitation Assessment Program (NAPAP). Case study prepared for the American Meteorological Society, 2002 Summer Policy Colloquium. http://www.ametsoc.org/atmospolicy/PolicyCaseStudies.html.

Holling, C.S. (ed.), 1978: *Adaptive Environmental Assessment and Management*. Wiley, New York, USA.

Holman, I.P., P.J. Loveland, R.J. Nicholls, S. Shackley, P.M. Berry, M.D.A. Rounsevell, E. Audsley, P.A. Harrison, and R. Wood, 2002: *REGIS - Regional Climate Change Impact Response Studies in East Anglia and North West England*. UK Climate Impacts Programme, UK.

Holtsmark, B.J. and K.H. Alfsen, 2005: PPP Correction of the IPCC emissions scenarios: does it matter? *Climatic Change* **68**, 1, 11-19.

Huss, W.R., 1988: A move toward scenario analysis. *International Journal of Forecasting* **4**, 377-388.

IPCC (Intergovernmental Panel on Climate Change), 1990: *Climate Change: The IPCC Scientific Assessment* [Houghton, J.T., G.J. Jenkins, and J.J. Ephraums (eds.)]. Cambridge University Press, Cambridge, UK.

IPCC (Intergovernmental Panel on Climate Change), 1996a: *Climate Change 1995: The Science of Climate Change* [Houghton, J.T., L.G. Meira Filho, B.A. Calander, N. Harris, A. Kattenberg, and K. Maskell (eds.)]. Cambridge University Press, Cambridge, UK.

IPCC (Intergovernmental Panel on Climate Change), 1996b: *Climate Change 1995: Economic and Social Dimensions of Climate Change* [Bruce, J.P., H. Lee, E.F. Haites (eds.)]. Cambridge University Press, Cambridge, UK.

IPCC (Intergovernmental Panel on Climate Change), 2001a: *Climate Change 2001, The Scientific Basis* [Houghton, J.T., Y. Ding, D.J. Griggs, M. Noguer, P.J. van der Linden, and D. Xiaosu. Cambridge University Press, Cambridge, UK.

IPCC (Intergovernmental Panel on Climate Change), 2001b: *Climate Change 2001: Impacts, Adaptations, and Vulnerability* [McCarthy, J.J., O.F. Canziani, N.A. Leary, D.J. Dokken, and K.S. White. Cambridge University Press, Cambridge, UK.

IPCC (Intergovernmental Panel on Climate Change), 2001c: *Climate Change 2001: Mitigation*, [Metz, B., O. Davidson, R. Swart, and J. Pan (eds.)]. Cambridge University Press, Cambridge, UK.

IPCC-TGCIA (Intergovernmental Panel on Climate Change-Task Group on Scenarios for Climate Impact Assessment), 1999: *Guidelines on the Use of Scenario Data for Climate Impact and Adaptation Assessment*, Version 1, [Carter, T.R., M. Hulme, and M. Lal (eds.)].

Johns, T.C., R. Carnell, J. Crossly et al., 1997: The second Hadley Centre Coupled Ocean-Atmosphere GCM: model description, spinup, and validation. *Climate Dynamics* **13**, 225-237.

Jones, W.M., 1985: *On Free-Form Gaming*. Rand Note N-2322-RC, Rand Corporation, Santa Monica, CA, USA.

Kahn, H. and A.J. Wiener, 1967: *The Year 2000: a Framework for Speculation on the Next Thirty-Three Years*. Macmillan, New York.

Kalkstein, L.S. and J.S. Greene, 1997: An evaluation of climate/mortality relationships in large U.S. cities and the possible impacts of a climate change. *Environmental Health Perspectives* **105**, 84–93

Keith, D.W., 2000: Geoengineering the climate: history and prospect. *Annual Review of Energy and Environment* **25**, 245-284.

Keith, D.W., M. Ha-Duong, and J.K. Stolaroff, 2006: Climate strategy with CO2 capture from the air. *Climatic Change* **74**, 1-3.

Kemp-Benedict, E., C. Heaps, and P. Raskin, 2002: Global Scenario Group Futures: Technical Notes. Stockholm Environment Institute PoleStar Series Report no. 9.

Kinney, P., et al., 2005: *Assessing Potential Public Health and Air Quality Impacts of Changing Climate and Land Use in Metropolitan New York.* Columbia Earth Institute, New York, USA.

Kinney, P.L., J.E. Rosenthal, C. Rosenzweig, C. Hogrefe, W. Solecki, K. Knowlton, C. Small, B. Lynn, K. Civerolo, J.Y. Ku, R. Goldberg, C. Oliveri, 2006: Assessing potential public health impacts of changing climate and land uses: the New York Climate & Health Project. In: *Climate Change and Variability: Impacts and Responses* [Ruth, M., K. Donaghy, and P. Kirshen (eds.)]. New Horizons in Regional Science, Edward Elgar, Cheltenham, UK.

Kittel, T.G.F., N.A. Rosenbloom, T.H. Painter, D.S. Schimel, and VEMAP Participants, 1995: the VEMAP Integrated Database for Modeling United States Ecosystem/Vegetation Sensitivity to Climate Change. *Journal of Biogeography* **22**, 4-5, 857-862.

Leggett, J., W. Pepper, R.J. Swart, J.A. Edmonds, L.G. Meira Filho, I. Mintzer, M.X. Wang, and J. Watson, 1992: Emissions scenarios for the IPCC, an update: assumptions, methodology, and results. In: *Climate Change 1992: The Supplementary Report to the IPCC Scientific Assessment* [Houghton, J.T., B.A. Callander, and S.K. Varney (eds.)]. Cambridge University Press, Cambridge, UK.

Lempert, R.J., D.G. Groves, S.W. Popper, and S.C. Bankes, 2006: A general, analytic method for generating robust strategies and narrative scenarios. *Management Science* **52**, 4, 514-528.

Levine, R.A., 1964a: Crisis games for adults. In: Crisis Games 27 Years Later: Plus C'est Déjà vu. P-7719, RAND Corporation, Santa Monica, CA, USA.

Levine, R.A., 1964b: Crisis games: a rejoinder to Tom Schelling and to some extent to Bill Jones. *Crisis Games 27 Years Later: Plus C'est Déjà vu.* P-7719, RAND Corporation, Santa Monica, CA, USA.

Levy, M.A., 1995: International co-operation to combat acid rain. In: *Green Globe Yearbook of International Co-operation on Environment and Development 1995* [Bergesen, H., G. Parmann, and Ø. Thommessen (eds.)]. Oxford University Press, Oxford, UK.

Lofgren, B.M., F.H. Quinn, A.H. Clites, R.A. Assel, and A.J. Eberhardt, 2000: Water resources. In: *Preparing for a Changing Climate: Great Lakes Overview.*

Lomborg, B. 2001: *The Skeptical Environmentalist.* Cambridge University Press, Cambridge, UK.

London Observer, 2004: Now the Pentagon tells Bush: climate change will destroy us. February 22:1. observer.guardian.co.uk/international/story/0,6903,1153513,00.html,

Lorenzoni, I., A. Jordan, M. Hulme, R.K. Turner, and T. O'Riordan, 2000: A co-evolutionary approach to climate change impact assessment: part i. integrating socio-economic and climate change scenarios. *Global Environmental Change* **10**, 1, 57-68.

MacCracken, M.C., E.J. Barron, D.R. Easterling, B.S. Felzer, and T.R. Karl, 2003: Climate change scenarios for the U.S. National Assessment. *Bulletin of the American Meteorological Society* **84**, 12, 1711-1723.

MacCracken, M.C., E. Barron, D. Easterling, B. Felzer, and T. Karl, 2001: Scenarios for climate variability and change. See NAST 2001.

McLean, R.F., A. Tsyban, V. Burkett, J. Codignotto, D. Forbes, V. Ittekkot, N. Mimura, and R.J. Beamish, 2001: Coastal zones and Marine ecosystems. In: *Climate Change: Impacts, Adaptation, and Vulnerability* [McCarthy, J.J., O.F. Canziani, N.A. Leary, D.J. Dokken, and K.S. White (eds.)]. Cambridge University Press, Cambridge, UK.

Manabe, S. and R.J. Stouffer, 1979: A CO2-climate sensitivity study with a mathematical model of the global climate. *Nature* **282**, 491-493.

Manabe, S., R.J. Stouffer, M.J. Spelman, and K. Bryan, 1991: Transient responses of a coupled ocean-atmosphere model to gradual changes of atmospheric CO2. part i: annual mean response. *Journal of Climate* **4**, 8, 785-818.

Manabe, S. and R.T. Wetherald, 1967: Thermal equilibrium of the atmosphere with a given distribution of relative humidity. *Journal of the Atmospheric Sciences* **24**, 3, 241-259.

Manne, A., R. Richels, and J.A. Edmonds, 2005: Market exchange rates or purchasing power parity: does the choice make any difference to the climate debate? *Climatic Change*, **71**, 1-2, 1-8.

MEA (Millennium Ecosystem Assessment), 2005: *Ecosystems and Human Well-Being: Synthesis Report.* Island Press, Washington, DC, USA.

MEA (Millennium Ecosystem Assessment), 2006: *Ecosystems and Human Well-Being: Scenarios.* Island Press, Washington, DC, USA.

Meadows, D.H., D.L. Meadows, J. Randers, W.W. Behrens III, 1972: *Limits to Growth: A Report to the Club of Rome.* Universe Books, New York, USA.

Mearns, L.O., C. Rosenzweig, and R. Goldberg, 1992: Effect of changes in interannual climatic variability on CERES-wheat yields: sensitivity and 2 x CO2 general circulation model studies. *Agricultural and Forest Meteorology* **62**, 159-189.

Mearns, L.O., C. Rosenzweig, and R. Goldberg, 1996: The effect of changes in daily and interannual climatic variability on CERES-wheat: a sensitivity study. *Climatic Change* **32**, 257-292.

Mearns, L.O., M. Hulme, T.R. Carter, R. Leemans, M. Lal, and P. Whetton, 2001: Climate scenario development. In: *Climate Change 2001: The Scientific Basis* [Houghton, J.T., Y. Ding, D.J. Griggs, M. Noguer, P.J. van der Linden, X. Dai, K. Maskell, and C.A. Johnson (eds.)]. Cambridge University Press, Cambridge, UK.

Michaels, P.J., 2003a: Science or political science: an assessment of the US National Assessment of the potential consequences of climate variability and change. In: *Politicizing Science: the Alchemy of Policymaking* [Gough, M. (ed.)]. Publication 517, Hoover Institution Press, Stanford, CA, USA.

Michaels, P. (ed.), 2003b: Bad math. *World Climate Report* **13**, 10.

Miller, S.S., 1990: NAPAP: a unique experience. *Environmental Science and Technology* **24**, **12**, 1781-1782.

Mitchell, J.F.B., S. Manabe, V. Meleshko, and V. Tokioka, 1990: Equilibrium climate change and its implications for the future. In: *Climate Change: The IPCC Scientific Assessment* [Houghton, J.T., G.J. Jenkins, and J.J. Ephraums (eds.)]. Cambridge University Press, Cambridge, UK.

Mitchell, R.B., W.C. Clark, D.W. Cash, and N.M. Dickson (eds.), 2006: *Global Environmental Assessments: Information and Influence*. MIT Press, Cambridge, MA, USA.

Morita, T. and H-C. Lee, 1998: IPCC SRES Database. *Database prepared for IPCC Special Report on Emissions Scenarios*. http:www-cger.nies.go.jp/cger-e/db/ipcc.html.

Morita, T., J. Robinson, A. Adegbulugbe, J. Alcamo, D. Herbert, E.L. La Rovere, N. Nakicenovic, H. Pitcher, P. Raskin, K. Riahi, A. Sankovski, V. Sokolov, B. de Vries, and Z. Dadi, 2001: Greenhouse gas emissions: mitigation scenarios and implications. In: *Climate Change 2001: Mitigation* [Metz, B., O. Davidson, R. Swart, and J. Pan (eds.)]. Cambridge University Press, Cambridge, UK.

Morgan, M.G. and D.W. Keith, 1995: Subjective judgments by climate experts. *Environmental Science & Technology* **29**, 468-476.

Morgan, M.G., M. Kandlikar, J. Risbey, and H. Dowlatabadi, 1998: Why conventional tools of policy analysis are often inadequate for problems of global change. *Climatic Change* **41**, 271-281.

Morgan, M.G., R. Cantor, W.C. Clark, A. Fisher, H.D. Jacoby, A.C. Janetos, A.P. Kinzig, J. Melillo, R.B. Street, and T.J. Wilbanks, 2005: Learning from the U.S. National Assessment of Climate Change. *Environmental Science & Technology* **39**, 9023-9032.

Morton, R.A., N.A. Buster, M.D. Krohn, 2002: Subsurface controls on historical subsidence rates and associated wetland loss in southcentral Louisiana. *Transactions - Gulf Coast Association of Geological Societies* **52**, 767-778.

Moss, R.H. and S.H. Schneider, 2000: Uncertainties in the IPCC TAR: recommendations to lead authors for more consistent assessment and reporting. In: Cross Cutting Issues Guidance Papers, Intergovernmental Panel on Climate Change, Geneva, Switzerland.

Mote, P.W., E.A. Parson, et al., 2003: Climate and the water, forests, and salmon of the Pacific Northwest. *Climatic Change* **61**, 1-2, 45-88.

Naki enovi, N. and R. Swart (eds.), 2000: Special Report on Emis-sions Scenarios. Cambridge University Press, Cambridge, UK. http://www.grida.no/climate/ipcc/emission/

Naki enovi, N., A. Grübler, S. Gaffin, T.T. Jung, T. Kram, T. Morita, H. Pitcher, K. Riahi, M. Schlesinger, P.R. Shukla, D. van Vuuren, G. Davis, L. Michaelis, R. Swart, and N. Victor, 2003: IPCC SRES revisited: a response. *Energy & Environment* **14**, 2-3, 187-214.

NAPAP (National Acid Precipitation Assessment Program, 1982: *Annual Report: National Acid Precipitation Assessment Program*. National Acid Precipitation Assessment Program, Washington, DC, USA.

Nash, J.F., 1950: Equilibrium points in n-person games. *Proceedings of the National Academy of Sciences* **36 (1)**, 48-49.

NAST (National Assessment Synthesis Team), 2000: *Climate Change Impacts on the United States: The Potential Consequences of Climate Variability and Change, Overview*. Cambridge University Press, Cambridge, UK.

NAST (National Assessment Synthesis Team), 2001: *Climate Change Impacts in the United States: Potential Consequences of Climate Change and Variability and Change*. Cambridge University Press, Cambridge, UK.

NRC (US National Research Council), 1996: *Understanding Risk: Informing Decisions in a Democratic Society*. [Stern, P.C. and H.V. Fineberg (eds.)]. Committee on Risk Characterization. National Academy Press, Washington, DC, USA.

NRC (US National Research Council), 1999: *Our Common Journey: A Transition toward Sustainability*. Board on Sustainable Development. National Academy Press, Washington, DC, USA.

NRC (US National Research Council), 2002: *Abrupt Climate Change: Inevitable Surprises.* Committee on Abrupt Climate Change. National Academy Press, Washington, DC, USA.

NRC (US National Research Council), 2005: *Decision Making for the Environment: Social and Behavioral Science Research Priorities.* Committee on Human Dimensions of Global Change. National Academy Press, Washington, DC, USA.

Parikh, J.K., 1992: IPCC strategies unfair to the South. *Nature* **360**, 507-508.

Parikh, J.K., 1998: The emperor's new clothes: long-range energy-use scenarios by IIASA-WEC and IPCC. *Energy* **23(1)**, 69-70.

Parson, E.A., 1996: What can you learn from a game? In: *Wise Choices: Games, Decisions, Negotiations* [Zeckhauser, R., R.L. Keeney, and J.K. Sebenius (eds.)]. Harvard Business School Press, Boston, USA.

Parson, E.A., 1997: Informing global environmental policy-making: a plea for new methods of assessment and synthesis. *Environmental Modeling and Assessment* **2(4)**, 267-279.

Parson, E.A., 2003: *Protecting the Ozone Layer: Science and Strategy.* Oxford University Press, New York, USA.

Parson, E.A., 2006: Reflections on air capture: the political economy of active intervention in the global environment. *Climatic Change* **74**, 1-3.

Parson, E.A. and K. Fisher-Vanden, 1997: Integrated assessment models of global climate change. *Annual Review of Energy and the Environment* **22**, 589-628.

Parson, E.A. and M.G. Morgan, with A. Janetos, L. Joyce, B. Miller, R. Richels, and T. Wilbanks, 2001: Socioeconomic context for climate impact assessment. In: *Climate Change Impacts on the United States*, NAST. Cambridge University Press, Cambridge, UK.

Parson, E.A., R.W. Corell, et al., 2003: Understanding climate impacts, vulnerabilities, and adaptation in the United States: building a capacity for assessment. *Climatic Change* **57(1)**, 9-42.

Payne, J.T., A.W. Wood, A.F. Hamlet, R.N. Palmer, and D.P. Lettenmaier, 2004: Mitigating the effects of climate change on the water resources of the Columbia River Basin. *Climatic Change* **62(1-3)**, 233-256.

Perhac, R.M., 1991: Usable science: lessons from acid rain legislation, NAPAP. *Power Engineering* **95**, 10, 26-29.

Pitcher, H.M., 2005. Downscaling: something for nothing? Presentation to workshop on global-change scenarios, Snowmass Colorado, July 26.

Pittock, A.B., R.N. Jones, and C.F. Mitchell, 2001: Probabilities will help us plan for climate change. *Nature* **413**, 249.

Providence Journal, 2004. Pentagon report plans for climate catastrophe. March 3, A1.

Raisanen, J. and T. Palmer, 2001: A probability and decision-model analysis of a multi-model ensemble of climate change simulations. *Journal of Climate* **14**, 3212-3226.

Raskin, P., T. Banuri, G. Gallopin, P. Gutman, A. Hammone, R. Kates, and R. Swart, 2002: *Great Transition: the Promise and Lure of the Times Ahead.* Global Scenario Group, Stockholm Environment Institute, Boston.

Reilly, J., P.H. Stone, et al., 2001: Climate change – uncertainty and climate change assessments. *Science* **293**, 430.

Renn, O., T. Webler, and P. Wiedemann, 1995: *Fairness and Compensation in Citizen Participation.* Kluwer, Dordrecht, The Netherlands.

Roberts, L., 1991: Learning from an acid rain program. *Science* **251**, 1302-1305.

Robinson, J.B., 1982: Energy backcasting: a proposed method of policy analysis. *Energy Policy* **10(4)**, 337-45.

Robinson, J.B., 2003: Future subjunctive: backcasting as social learning. *Futures* **35**, 839-856.

Rosenberg, N., P. Crosson, K. Frederick, W. Easterling, M. McKenney, M. Bowes, R. Sedjo, J. Darmstadler, L. Katz, and K. Lemon, 1993: The MINK methodology: background and baseline. *Climatic Change* **24**, 7-22.

Rosenzweig, C. and W.D. Solecki (eds.), 2001: Climate change and a global city: the potential consequences of climate variability and change - metro east coast. Columbia Earth Institute, New York, USA.

Rosenzweig, C. and D.C. Major, 2006: *Climate Impact Assessment of Environmental Infrastructure Systems: Phase I Final Report: Scoping for Phase II.* Center for Climate Systems Research, Columbia University, New York, USA.

Rothman, D., J. Robinson, and D. Biggs, 2003: Signs of life: linking indicators and models in the context of QUEST. In: *Implementing Sustainable Development - Integrated Assessment and Participatory Decision-Making Processes* [Abaza, H. and A. Baranzini (eds.)]. Edward Elgar, Cheltenham, UK.

Rubin, E.S., 1991: Benefit-cost implications of acid rain controls: an evaluation of the NAPAP integrated assessment. *Journal of the Air and Waste Management Association* **41(7)**, 914-921.

Ruosteenoja, K., T.R. Carter, K. Jylha, and H. Tuomenvirta, 2003: *Future Climate in World Regions: An Intercomparison of Model-Based Projections for the New IPCC Emissions Scenarios.* Finnish Environment Institute, Helsinki.

Russell, M., 1992: Lessons from NAPAP. *Ecological Applications* **2(2)**, 107-110.

San Francisco Chronicle, 2004. Pentagon-sponsored climate report sparks hullabaloo in Europe. February 25, A5.

Schelling, T.C., 1964: An uninhibited pitch for crisis games. In: *Crisis Games 27 Years Later: Plus C'est Deja* Vul. P-7719, RAND Corporation, Santa Monica, CA, USA.

Schelling, T.C., 1994: Nuclear history program oral history transcript. NHP Berlin Crisis Oral History Project, Interview with Thomas Schelling. University of Maryland Center for International Security Studies, Nuclear History Program, College Park, MD, USA.

Schelling, T.C., 1983: Climate change: implications for welfare and policy. In: *Changing Climate*, Report of the Carbon Dioxide Assessment Committee, US National Research Council. National Academy Press, Washington DC, USA.

Schneider, S.H., 2001: What is "dangerous" climate change? *Nature* **411**, 17-19.

Schneider, S.H., 2002: Can we estimate the likelihood of climatic changes at 2100? *Climatic Change* **52(4)**, 441-451.

Schoemaker, P.J.H., 1995: Scenario planning: a tool for strategic thinking. *Sloan Management Review*, Winter.

Schultz, R.L. and E.M. Sullivan, 1972: Developments in simulation in social and administrative science. In: *Simulation in the Social and Administrative Sciences* [Guetzkow, H., P. Kotler, and R.L. Schultz (eds.)]. Prentice-Hall, Englewood Cliffs, NJ, USA.

Schwartz, P., 1991: *The Art of the Long View: Planning for the Future in an Uncertain World.* Currency Doubleday, New York, USA.

Semenov, M.A. and J.R. Porter, 1995: Climatic variability and the modeling of crop yields. *Agricultural and Forest Meteorology* **73**, 265-283.

Shell International, 2001: *Energy Needs, Choices, and Possibilities: Scenarios to 2050.* Global Business Environment.

Shell International, 2003: *Scenarios: an Explorer's Guide. Global Business Environment.* www-static.shell.com/static/royal-en/downloads/scenarios_explorersguide.pdf.

Shubik, M., 1975: *The Uses and Methods of Gaming.* Elsevier, New York, USA.

Shinkle, K.D. and R.K. Dokka, 2004: *Rates of Vertical Displacement at Benchmarks in the Lower Mississippi Valley and the Northern Gulf Coast.* NOAA Technical Report NOS/NGS 50, US Department of Commerce, National Oceanic and Atmospheric Administration, Washington, DC, USA.

Smil, V., 2005: *Energy at the Cross Roads.* MIT Press, Cambridge, MA, USA.

Southeast Regional Assessment Team, 2002: *Preparing for a Changing Climate: Potential Consequences of Climate Variability and Change, Southeast.* Publication 8-40002, Global Hydrology and Climate Center, University of Alabama in Huntsville, USA.

Stipp, D., 2004: The Pentagon's weather nightmare: the climate could change radically, and fast. *Fortune* February 9, 100.

Stockholm Environment Institute, 1999: *POLESTAR: System Manual for Version 2000.* POLESTAR Series Report No. 2., Stockholm Environment Institute, Sweden.

Stouffer, R.J., S. Manabe, and K. Bryan, 1989: Interhemispheric asymmetry in climate response to a gradual increase of CO_2. *Nature* **342**, 660-662.

Svedin, U. and B. Aniansson (eds.), 1987: *Surprising Futures.* Notes from an International Workshop on Long-term World Development, Friibergh Manor, Sweden, January 1986. Report 87:1, Swedish Council for Planning and Coordination of Research, Stockholm.

Swart, R.J., P. Raskin, and J. Ribonson, 2004: The problem of the future: sustainability science and scenario analysis. *Global Environmental Change* **14**, 137-146.

Tebaldi, C., L.O. Mearns, R.L. Smith, and D. Nychka, 2004: Regional probabilities of precipitation change: a Bayesian analysis of multimodel simulations. *Geophysical Research Letters* **31**, L24213.

Tebaldi, C., R.L. Smith, D. Nychka, and L.O. Mearns, 2005: Quantifying uncertainty in projections of regional climate change: a Bayesian approach to the analysis of multimodel ensembles. *Journal of Climate* **18(10)**, 1524-1540.

Terleckyj, N.E., 1999a: *Analytic Documentation of Three Alternate Socioeconomic Projections, 1997-2050.* NPA Data Services, Washington, DC, USA.

Terleckyj, N.E., 1999b: *Development of Three Alternate National Projection Scenarios, 1997-2050.* NPA Data Services, Washington, DC, USA.

Toth, F. and T.W. Wilbanks, 2004: *Considering the Technical and Socioeconomic Assumptions Embedded in the SRES Scenario Families*. Guidance Paper, Fourth Assessment Report, IPCC Working Group 2. September 2004.

Tversky, A. and D. Kahneman, 1974: Judgment under uncertainty: heuristics and biases. *Science* **185**, 1124-1131.

UKCIP (UK Climate Impacts Programme), 1998: *Climate Change Scenarios for the United Kingdom*. UKCIP Technical Report No. 1, Oxford, UK. http://www.ukcip.org.uk/resources/publications/pub_dets.asp?ID=11.

UKCIP (UK Climate Impacts Programme), 2000: *Climate Change: Assessing the Impacts, Identifying the Responses* [McKenzie Hedger, M., M. Gawith, I. Brown, R. Connell, and T.E. Downing (eds.)]. UKCIP Short Report, Oxford, UK. http://www.ukcip.org.uk/resources/publications/pub_dets.asp?ID=16.

UKCIP (UK Climate Impacts Programme), 2001: *Socio-Economic Scenarios for Climate Change Impact Assessment: A Guide to Their Use in the UKCIP*. UKCIP, Oxford, UK. ukcip.org.uk/resources/publications/documents/34.pdf.

UKCIP (UK Climate Impacts Programme), 2005: *Measuring Progress: Preparing for Climate Change through the UK Climate Impacts Programme* [West, C. and M. Gawith (eds.)]. UKCIP, Oxford, UK. http://www.ukcip.org.uk/resources/publications/pub_dets.asp?ID=68.

UK House of Lords, 2005: *The Economics of Climate Change*. Select Committee on Economic Affairs. 6 July. The Stationery Office Limited, London, UK.

UK Office of Science and Technology, 2002: *Foresight Futures 2020: Revised Scenarios and Guidance*. Department of Trade and Industry, London, UK.

US Army Corps of Engineers, 2004: *Louisiana Coastal Area (LCA), Louisiana, Ecosystem Restoration Study*. New Orleans District, Louisiana, USA.

US Climate Change Science Program and Subcommittee on Global Change Research, 2003: *Strategic Plan for the U.S. Climate Change Science Program*. Washington, DC, USA. www.climate-science.gov/Library/stratplan2003/final/default.htm.

US EPA (US Environmental Protection Agency), 1989: *The Potential Effects of Global Climate Change on the United States* [Smith, J. and D. Tirpak (eds.)]. EPA-230-05-89-050, Washington, DC, USA.

US OTA (US Congress, Office of Technology Assessment), 1993: *Preparing for an Uncertain Climate* (2 vols.). OTA-O-567 and -568, US Government Printing Office, Washington, DC, USA.

Van der Heijden, K., 1996: *Scenarios: The Art of Strategic Conversation*. John Wiley and Sons, Chichester, UK.

van Notten, P.W.F., J. Rotmans, M.B.A. van Asselt, and D.S. Rothman, 2003: An updated scenario typology. *Futures* **35**, 423–443.

van't Klooster, S.A. and M.B.A. van Asselt, 2006: Practicing the scenario-axes technique. *Futures* **38**, 15-30.

Vaughan, D.G. and J.R. Spouge, 2002: Risk estimation of collapse of the West Antarctic Ice Sheet. *Climatic Change* **52**, 65-91.

VEMAP Members, 1995: Vegetation/Ecosystem Modeling and Analysis Project (VEMAP): comparing biogeography and biogeochemistry models in a continental-scale study of terrestrial ecosystem responses to climate change and CO2 doubling. *Global Biogeochemical Cycles* **9(4)**, 407-437.

von Neumann, J. and O. Morgenstern, 1944: *Theory of Games and Economic Behavior*. Princeton University Press, Princeton, NJ, USA.

Wack, P., 1985a: Scenarios: uncharted waters ahead. *Harvard Business Review* **63(5)**, 73-89.

Wack, P., 1985b: Scenarios: shooting the rapids. *Harvard Business Review* **63(6)**, 139-150.

Wallsten, T.S. and R.G. Whitfield, 1986: *Assessing the Risks to Young Children of Three Effects Associated with Elevated Blood Lead Levels*. ANL/AA-32, Argonne National Laboratory, Argonne, IL, USA.

Washington, W.M. and G.A. Meehl, 1989: Climate sensitivity due to increased CO2: experiments with a coupled atmosphere-ocean general circulation model. *Climate Dynamics* **4(1)**,1-38.

WBCSD (World Business Council on Sustainable Development), 1997: *Exploring Sustainable Development: WBCSD Global Scenarios, 2000-2050*. www.wbcsd.org/DocRoot/FFiAJwjBGGNjlawOAipD/exploringscenarios.pdf.

Webster, M.D., 2003: Communicating climate change uncertainty to policy-makers and the public. *Climatic Change* **61**, 1-2, 1-8.

WEC/IIASA (World Energy Council, International Institute for Applied Systems Analysis), 1995: *Global Energy Perspectives to 2050 and Beyond*. World Energy Council, London, UK.

Weyant, J., O. Davidson, H. Dowlatabadi, J.A. Edmonds, M. Grubb, E.A. Parson, R. Richels, J. Rotmans, P.R. Shukla, R.S.J. Tol, W. Cline, and S. Fankhauser, 1996: Integrated assessment of climate change: an overview and comparison of approaches and results. In: *Climate Change 1995, Economic and Social Dimensions of Climate Change* [Bruce, J.P., H. Lee, and E.F. Haites (eds)]. Cambridge University Press, Cambridge, UK.

Weyant, J.P. and J.N. Hill, 1999: Introduction and overview, the costs of the Kyoto Protocol: a multi-model evaluation. *The Energy Journal*, Special Issue, vii-xliv.

Weyant, J.P. (ed.), 2004: EMF 19: alternative technology strategies for climate change policy. *Energy Economics* **26**, 4.

Wigley, T.M.L., R. Richels, and J.A. Edmonds, 1996: Economic and environmental choices in the stabilization of atmospheric CO2 concentrations. *Nature* **379**, 240-243.

Wilby, R.L. and T.M.L. Wigley, 1997: Downscaling general circulation model output: a review of methods and limitations. *Progress in Physical Geography* **21**, 530-548.

Williams, K.L., K.C. Ewel, R.P. Stumpf, F.E. Putz, and T.W. Workman, 1999: Sea-level rise and coastal forest retreat on the west coast of Florida. *Ecology* **80(6)**, 2045-2063.

CONTACT INFORMATION

Global Change Research Information Office
c/o Climate Change Science Program Office
1717 Pennsylvania Avenue, NW
Suite 250
Washington, DC 20006
202-223-6262 (voice)
202-223-3065 (fax)

The Climate Change Science Program incorporates
the U.S. Global Change Research Program and the
Climate Change Research Initiative.

To obtain a copy of this document, place an order at
the Global Change Research Information Office
(GCRIO) web site: http://www.gcrio.org/orders

CLIMATE CHANGE SCIENCE PROGRAM AND THE SUBCOMMITTEE ON GLOBAL CHANGE RESEARCH

William J. Brennan, Chair
Department of Commerce
National Oceanic and Atmospheric Administration
Acting Director, Climate Change Science Program

Jack Kaye, Vice Chair
National Aeronautics and Space Administration

Allen Dearry
Department of Health and Human Services

Jerry Elwood
Department of Energy

Mary Glackin
National Oceanic and Atmospheric Administration

Patricia Gruber
Department of Defense

William Hohenstein
Department of Agriculture

Linda Lawson
Department of Transportation

Mark Myers
U.S. Geological Survey

Jarvis Moyers
National Science Foundation

Patrick Neale
Smithsonian Institution

Jacqueline Schafer
U.S. Agency for International Development

Joel Scheraga
Environmental Protection Agency

Harlan Watson
Department of State

EXECUTIVE OFFICE AND OTHER LIAISONS

Melissa Brandt
Office of Management and Budget

Stephen Eule
Department of Energy
Director, Climate Change Technology Program

Katharine Gebbie
National Institute of Standards & Technology

Margaret McCalla
Office of the Federal Coordinator for Meteorology

George Banks
Council on Environmental Quality

Gene Whitney
Office of Science and Technology Policy

www.ingramcontent.com/pod-product-compliance
Lightning Source LLC
Chambersburg PA
CBHW081502170526
45166CB00008B/2525